Introduction to Nonlinear Optics of Photonic Crystals and Metamaterials (Second Edition)

Introduction to Nonlinear Optics of Photonic Crystals and Metamaterials (Second Edition)

Arthur R McGurn
Department of Physics, Western Michigan University, Kalamazoo, MI, USA

IOP Publishing, Bristol, UK

ISBN 978-0-7503-3579-9 (ebook)
ISBN 978-0-7503-3577-5 (print)
ISBN 978-0-7503-3580-5 (myPrint)
ISBN 978-0-7503-3578-2 (mobi)

DOI 10.1088/978-0-7503-3579-9

Version: 20211101

IOP ebooks

British Library Cataloguing-in-Publication Data: A catalogue record for this book is available from the British Library.

Published by IOP Publishing, wholly owned by The Institute of Physics, London

IOP Publishing, Temple Circus, Temple Way, Bristol, BS1 6HG, UK

US Office: IOP Publishing, Inc., 190 North Independence Mall West, Suite 601, Philadelphia, PA 19106, USA

To my wife, Maria

Contents

Preface

This book is pedagogical. It is meant as an introduction to the ideas and phenomena that occur in the nonlinear optics of photonic crystals and metamaterials. At the same time it also provides the student with a basic introduction to the general ideas of nonlinear optics. The book should not, however, be viewed as a comprehensive review of the topics of photonic crystals and metamaterials and their nonlinear properties but rather as a survey of some of the most important fundamentals of the field. In this regard, it provides a strong background to a student new to the field. The phenomena discussed are illustrated within the context of simple models, offering an easy understanding of the physical phenomena that are important in these two rapidly developing areas of nanophotonic technology.

The book first deals with nonlinear dielectric properties. A model based on classical physics is used to develop some of the ideas of the Kerr nonlinear dielectric and the generation of second and higher harmonics of radiation in a semi-quantitative way. The treatment is similar to that used in basic electrodynamics texts to provide an understanding of the dielectric properties of linear dielectric materials. An emphasis is on nonlinear effects as being small and on their treatment within the context of them entering the considerations as perturbations of otherwise linear systems. Consequently, along with the development of the dielectric properties, the theory of multiple scale perturbation theory is simultaneously formulated as an important technique in correctly dealing with the physics of systems with small nonlinearities. The reasons why multiple scale perturbation techniques are needed and a number of examples of the multiple scale treatment of nonlinear systems are given.

The distinctions between the ideas of photonic crystals and metamaterials and the different types of applications they address are discussed with emphasis on their elementary properties. Photonic crystals have dielectric properties which exhibit periodicity in space. Due to their spatial periodicity the dispersion relation of light within the photonic crystal exhibits a sequence of frequency pass bands and stop bands. Light at pass band frequencies propagates in the photonic crystal and light at stop band frequencies does not propagate through the photonic crystal. On the other hand, metamaterials are composed as arrays of nanoscopic electromagnetic resonator units. They are engineered to appear to be homogeneous media on the scale of the wavelengths of light with which they are designed to interact. The frequency dependence of the resonators allow metamaterials to exhibit frequency dependent refractive indices not otherwise observed in nature. Of particular importance is their ability to display negative index of refraction.

The theory of photonic crystals is greatly influenced by their spatial periodicity. In this regard, basic discussions of the dynamics of systems defined on a periodic lattice are made with a focus on the Maxwell equations expressed for periodic media. The direct and reciprocal lattices, Brillouin zones, Block functions, and the properties of stop and pass bands are introduced at an elementary level. Waveguides and single site impurities introduced into otherwise periodic bulk photonic crystal systems also are discussed at an introductory level, as are the techniques of Wannier functions and computer simulation treatments applied to these types of problems.

The discussions of periodic systems are also extended to considerations of the basic theory of surface gratings and applications of Bragg gratings within fiber optics and general optical waveguides. The use of photonic crystal technology in the design of optical fibers is an additional consideration of new designs based on the ideas of periodic media. In this regard, periodicity is introduced both as a novel means of making fiber claddings and as a confining mechanism based on stop bands. In addition, discussions of soliton modes within optical fibers formed of nonlinear optical media is briefly presented, offering a comparison with the soliton-like intrinsic localized modes present in photonic crystal waveguides formed of nonlinear optical media.

The application of metamaterials in the design of media with unusual refractive properties is discussed, with a particular focus on the development of materials exhibiting a negative refractive index. Metamaterials are typically designed as arrays of electromagnetic nanoresonators, and a commonly used resonator form is that of a split ring resonator (SRRs). In this regard, an understanding of the resonant properties of SRRs is developed in a detailed but elementary treatment. How these resonant properties are then employed to engineer a material exhibiting negative refraction is then explained in detailed.

The refractive properties of plane waves traveling between semi-infinite media of positive and negative refractive index media are studied comprehensively. Differences in the radiative properties of antennas and radiating charges within both positive and negative index media are compared. The essential operating principles of perfect lenses are explained and a comparison is made with the properties of lenses formed only of positive refractive index materials. In addition, the application of metamaterials in the formulation of the ideas of optical cloaking mechanisms are developed.

Both analytic and computer simulation methods are introduced at an elementary level for the treatment of basic problems in photonic crystals and metamaterials. In terms of computer solutions, the plane wave method, the finite difference time domain method, the method of moments, the conjugate gradient method, and the finite element method of computer simulation techniques are formulated and discussed in elementary and detailed manners. Various formulations for the boundary conditions of computer simulation studies needed to accurately approximate the interaction of radiation with dielectric features in infinite systems are presented.

On the side of analytic models, a number of difference equation formulations are given for photonic crystal and metamaterial waveguides. These difference equation models are used to facilitate the understanding of the basic physics of photonic crystals and metamaterials. The difference equation models are also realizable in many experimentally accessible cases. They often provide simple exactly solvable models whose properties are easily made transparent and are also representative of more complex general systems studied using computer simulations. Analytic techniques for the solution of difference equation formulations are discussed and are related to the physics of many different types of physical systems that occur in physics and technology.

The properties of photonic crystal and metamaterial systems with Kerr nonlinearity are discussed in terms of formulations based on difference equation models. Properties of optical bistability and resonant transmission anomalies are explained in detail based on elementary treatments in the difference equation formulations. The various properties are treated with discussions for their applications in devices and examples given in the context of the simple models.

A chapter is dedicated to discussions of computer simulation studies of nonlinear properties of some recently considered systems and the properties with which these have dealt. One-dimensional photonic crystals formed of a periodic layering of Kerr slabs are discussed along with the properties of Kerr impurities in higher-dimensional photonic crystals. In another important treatment, studies of the applications of photonic crystals to enhance the generation of second harmonics are outlined as another application of the use of photonic crystals. Some recent computer simulation studies of nonlinear metamaterials and the properties these studies have treated are also summarized. In these discussions only some of the important basic results from computer simulations are presented, and the discussions are not meant to be a comprehensive review of the field of computer simulation applied to nonlinear systems.

In the last chapter of the book a treatment is given of the effects of impurities and disorder introduced into photonic crystals and metamaterials. Single site impurities and finite clusters of impurities are discussed in the context of Green's functions theories and with the applications of ideas of group theory. General random disorder is treated through the application of the coherent potentional approximation, and the theory of this methodology is explained in detail. The general theory of Anderson localization in random disordered optical media is discussed along with the ideas of a scaling theory approach to the effects of disorder on the transport properties of disordered systems. In this regard, optical systems have provided important systems for the discussion of the ideas of so-called strong and weak localization. As a final treatment of localization, a focus is given on the effects of weak localization observed in the general diffuse elastic scattering of light and the general diffusely generated second harmonics of light from a randomly rough surface which supports surface electromagnetic waves. In this context, an introductory treatment is presented to the theory of surface electromagnetic waves and the requirements for their existence on a rough interface.

As a final note: both the Gaussian and Standard International Units formulations of the Maxwell equations are used at different points in the text. This is done as many of the results referenced from the literature occur in these sets of units, as will be indicated in the text. In addition, the author would like to thank the Department of Physics, University of California, Riverside, for allowing me the use of the University Library during the course of this project.

Arthur R McGurn
Rancho Mirage, California
May, 2021

Acknowledgement

The author wishes to thank Western Michigan University for the opportunity to write this book and the Department of Physics and Astronomy, University of California, Riverside, for the extension of library privileges at the UCR Library. The author also expresses his appreciation for former students, all of whom have pursued successful careers in physics and engineering.

Author biography

Arthur R McGurn

Professor Emeritus Arthur R McGurn, CPhys, FInstP, is a Fellow of the Institute of Physics, a Fellow of the American Physical Society, a Fellow of the Optical Society of America, a Fellow of the Electromagnetics Academy, and an Outstanding Referee for the journals of the American Physical Society. He received the PhD in Physics in 1975 from the University of California, Santa Barbara, followed by postdoctoral studies at Temple University, Michigan State University, and George Washington University (NASA Langley Research Center). The research interests of Professor McGurn have included works in the theory of: magnetism in disorder materials, electron conductivity, the properties of phonons, ferroelectrics and their nonlinear dynamics, Anderson localization, amorphous materials, the scattering of light from disordered media and rough surfaces, the properties of speckle correlations of light, quantum optics, nonlinear optics, the dynamical properties of nonlinear systems, photonic crystals, and metamaterials. He has over 150 publications spread amongst these various topics. Since 1981, he has taught physics at Western Michigan University, where he is currently a Professor of Physics and a WMU Distinguished Faculty Scholar. A number of PhD students have graduated from Western Michigan University under his supervision.

Symbols

ε	Electric permittivity
μ	Magnetic permeability
λ	Wavelength (m)
k	Wavevector
P	Charge density
B	Magnetic induction
H	Magnetic field
E	Electric field
D	Electric displacement
n	Index of refraction
L	Inductance
C	Capacitance
ω	Angular frequency
υ	Frequency
S	Poynting vector

Chapter 1

Introduction

In this book an introduction and discussion of some of the basic principles of linear and nonlinear optical nano-systems are given. The focus is on engineered optical systems that have been of recent interest in physics, engineering, and applied mathematics for their opto-electronic applications. These include photonic crystals and metamaterials, and in the following discussions the operating principles of photonic crystals and metamaterials are outlined. Photonic crystals, which have been of great interest in opto-electronic designs, are materials that exhibit a periodic dielectric variation in space and are designed to manipulate light with wavelengths of order of the length scales of the periodicity of the photonic crystal lattice [1–5]. The manipulation of light is accomplished through the use of the Bragg scattering properties of the periodic lattice of the dielectric which may be periodic in one, two, or three dimensions. Metamaterials often have periodic dielectric and magnetic properties but are designed to manipulate light of wavelengths much greater than the periodicity of the dielectric and magnetic lattice [1–5]. Metamaterials are designed to exhibit homogeneous properties to the light traveling within them. While photonic crystals are of interested for the diffractive effects they have on light propagating within them, metamaterials are of interest for the permittivity and permeability that they exhibit when they are treated as a uniform medium.

For both systems a particular goal of our presentation is to describe their behaviors as nonlinear optical systems [6–9]. To accomplish this, a general review is also given of the properties of these systems considered as linear optical systems. The treatments assume an undergraduate background in electrodynamics and are developed starting from an elementary level. The presentation provides an exposition of the basic principles of the photonic crystal and metamaterial systems. The wave excitations found in linear medium systems and their elementary properties are treated in detail. A detailed development, requiring no previous background in nonlinear materials, of the excitations in nonlinear systems consisting of plane waves and bright, dark, and gray solitons is also given.

1.1 Photonic crystals

Photonic crystals are periodic arrays of different dielectric materials [1–5]. They are literally crystals formed from dielectric materials rather than the individual atoms of materials such as crystalline NaCl, $CaCl_2$, etc The interest in the dielectric array is that it exhibits a band structure for the propagation of electromagnetic modes similar to the electronic band structure exhibited in metallic and semiconductor materials used in electronic designs [10]. As with the electron systems, the interest in the optical systems is in modes that have wavelengths of order of the length scale of the periodicity of the system. At these length scales the system displays a variety of diffraction effects on the excitations in the system. Various pass and stop frequency bands are opened in the frequency spectrum of excitations that can exist in the photonic crystal. Light at frequencies in a pass band will propagate through the photonic crystal while light at frequencies in a stop band will not propagate in a photonic crystal and are typically reflected from the system.

The existence of pass and stop frequency bands is the basis for engineering applications of photonic crystals. A cavity resonator, for example, can be formed as a cavity within the bulk of a photonic crystal. An electromagnetic wave with a frequency in the stop band of the photonic crystal cannot propagate out of the cavity through the bulk photonic crystal so it will be confined to the cavity. Such cavities offer high Q resonance cavities for laser applications that are not available through other technologies [11]. Another application of photonic crystal cavities is in the suppression and enhancement of atomic transitions. If the excited state of an atom which radiates at a stop band frequency is put in the cavity it will be suspended in the excited state due to its inability to radiate the excitation energy though the bulk of the photonic crystal. Similarly, the frequency mode density of states in a photonic crystal pass band can be enhanced from those of free space. This enhancement of the density of states increases the rate of decay of an excited atom within a photonic crystal cavity into enhanced pass band frequency modes over its free space decay rate [11].

The stop band effect can also be used in the design of waveguides meant to channel the flow of light through space [1–5]. A channel cut through the bulk of a photonic crystal will restrict light at frequencies in the stop bands of the bulk photonic crystal from propagating away from the waveguide, i.e., light at stop band frequencies will only move along the channel of the waveguide. Waveguides based on such photonic crystal designs can be more effective than fiber optic technologies in forming optical circuits. For example photonic crystal waveguides afford the possibility of sharper bends in their guiding channels and lower losses in general than are found in traditional fiber optics approaches [12]. In addition, photonic crystals can be designed from a wider variety of materials than are typically used in fiber optics.

One method of introducing a waveguide into a two-dimensional photonic crystal formed as a periodic array of identical parallel axis dielectric cylinders will be a focus in later considerations of guided modes. For this system the interest is in light propagating in the plane perpendicular the axes of the cylinders. A waveguide is introduced into the cylinder array by replacing a periodic array of dielectric cylinders of the photonic crystals by a set of identical impurity cylinders with different dielectric properties from those of the photonic crystal. This is done along a crystal axis of the two-dimensional array of the photonic crystals, and by choosing the impurity dielectric correctly a set of waveguide modes bound to and propagating along the array of waveguide impurity cylinders can be made to exist at stop band frequencies of the original photonic crystal. Later it is shown that the modes of this kind of waveguide system can be represented by a set of difference equations. This representation provides a helpful means of understanding the physics of photonic crystal waveguides and circuits.

One of the earliest applications of the ideas of photonic crystals, occurring in the initial stages of the developments of laser technology, was the application of one-dimensional or layered photonic crystals in the design of laser mirrors [1–5, 13]. Here the periodic layering of different dielectric media can be used to create low loss, highly reflective mirrors [13]. Similar layering effects are observed in insects in which the metallic, mirrored appearance of the shell of the animal is due to the layering of dielectrics rather than to a presence of metallic reflecting elements in the shell of the insect. In both of these systems the presence of pass and stop bands of the layering leads to the functioning of the reflecting surfaces. In addition to the insect example of photonic crystals developed in nature, a number of periodic one-, two-, and three-dimensional nano-systems that occur naturally in plant and animal materials have been suggested as a basis for the design of photonic crystals [1–5, 13]. The interest in these comes from the pre-existing nano-scale periodicity which is hard to create in a laboratory. A number of laboratory techniques have been developed for creating photonic crystals, but these are not discussed here.

In addition to photonic crystals formed as arrays of slabs, or cylinders, or three-dimensional periodic ordering of dielectric features, surfaces with periodic surface profiles are also found to exhibit important photonic crystal properties [1–5]. Periodic surfaces have long been studied for their applications as diffraction gratings, and a brief review of their basic theory is presented. In addition, under certain well defined conditions periodic surfaces can support surface electromagnetic waves. These are electromagnetic modes which are bound to the interface between two media, representing excitations localized about the interface and propagating parallel to the interface. These surface waves are important excitations with applications in electromagnetic scattering from the surface, in the design of various sensors, and in surface enhanced Raman scattering (SER) [1–5]. They feature prominently in the diffuse scattering of light at rough surfaces and in the diffuse generation of second harmonics of light at rough surfaces.

In regards to nonlinear effects in photonic crystals and waveguides, a treatment of the basic properties of parametric oscillators and parametric amplifiers is given [6, 7]. These are discussed in both bulk optical media and in the context of nonlinear fiber optical waveguides [12]. Bragg grating technology in fiber optics is also a topic. These systems are represented in many important technological applications.

1.2 Metamaterials

Metamaterials are artificial materials that are designed as arrays of nano-circuits known as split ring resonators (SRRs) [14]. SRR nano-circuits are inductor–capacitor resonator circuits which in metamaterial designs may be arrayed on a periodic nano-lattice or in a more complex array arrangement. The resonator circuits are known as SRRs because in the basic form of their design they are a metal ring with a split gap interrupting the ring. The loop of the ring provides self-inductance to the SRR, and the gap of the SRR is filled with a dielectric material to form a capacitor. Composed in this way the SRR is found to resonate at the frequency $\omega_{\mathrm{SRR}} = \frac{1}{\sqrt{LC}}$, where L is the SRR self-inductance and C is its capacitance [15–21]. For practical applications more complex SRR designs are used but all of the various designs operate on the basic principles that are outlined above. Within a particular metamaterial array the SRRs are coupled together by mutual inductive couplings between the various SRRs. This gives rise to many-body effects which also affect the optics of the material.

The purpose behind the composition of metamaterials as arrays of SRRs is to use their flexibility of design to create materials with enlarged sets of permittivity, ε, and permeability, μ, from those of naturally occurring materials [15–24]. In naturally occurring molecular solids the permittivity and permeability of the bulk solid arises from those of the molecules forming the systems. These properties are fundamentally limited by the nature and interactions of the molecular constituents of the materials. For example, in no naturally occurring material is it found that $\varepsilon < 0$ and $\mu < 0$ simultaneously at a single frequency of an applied electromagnetic plane wave. Regions of frequency are often found for which $\varepsilon < 0$ and regions of frequency are often found for which $\mu < 0$, but no natural materials have been found for which these two conditions are simultaneously satisfied at the same frequency. This limitation on ε and μ comes from the size restrictions of the molecular units and their effect on the magnetic response of molecular solids. The development of an artificial nano-circuit array allows these restrictions to be overcome.

At microwave and terahertz frequencies engineered SRR materials with $\varepsilon < 0$ and $\mu < 0$ simultaneously can be formulated at these frequencies through the use of an SRR array to customize the magnetic response of the material. The SRR ring geometry is arranged so that the resonant frequency $\omega_{\mathrm{SRR}} = \frac{1}{\sqrt{LC}}$ of the ring is in the microwave or terahertz region. Under this condition it is found that the size of the SRRs needed is much greater than the molecules available in molecular solids. In addition, at frequencies near ω_{SRR} the wavelength of the externally applied electromagnetic waves is slowly varying over a volume containing many SRRs so that the SRR material appears to the external wave to be a continuous medium. As the frequency, ω, of an externally applied field is changed so as to pass through the SRR's resonance at ω_{SRR}, the magnetic moment of the SRR changes sign. Consequently, about the resonance of the SRR system the SRR can exhibit a frequency region in which $\mu < 0$. This artificially induced $\mu < 0$ is easily correlated with a region of $\varepsilon < 0$ for the dielectric media in which the rings are embedded. The resulting metamaterial displays a set of frequencies for which $\varepsilon < 0$ and $\mu < 0$ at the same time.

1.3 Negative index of refraction

With the introduction of metamaterials a full range of ε and μ values become available for optical design. In particular, the case of $\varepsilon < 0$ and $\mu < 0$ is now a practical possibility. This is very important for refractive optics as materials with $\varepsilon < 0$ and $\mu < 0$ will exhibit negative refractive indices [18, 22, 25, 26]. The possibility of negative indexed materials was theoretically considered early in the twentieth century but was only treated as a curiosity [25, 26]. In naturally occurring positive index materials figure 1.1 shows that light incident on a planar interface from a positive indexed medium to an optically rarer or denser positive indexed medium is refracted from the second to the fourth quadrants. In negative index materials figure 1.2 shows that light incident on a planar interface from a positive indexed medium to a negative indexed material is refracted from the second to the third quadrants. Consequently, light incident in the second quadrant can now be guided though any angle in the forward direction of the third and fourth quadrants. This has found applications in cloaking protocols in which light directed at an object can be guided around it by using a continuous variation of positive and negative refractive indices [14, 18, 27]. Once around the object the light is sent along a path set to give the appearance that the object is not present. The object is made to appear not present even in its presence. Other such variations of positive and negative indices have found applications in simulating optical effects from general relativity. In noting these interesting properties of metamaterials, it should also be pointed out that under special circumstances the Bragg reflection properties of photonic crystals,

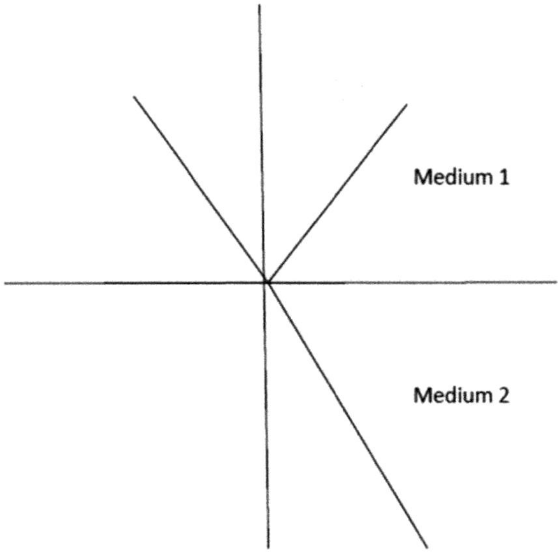

Figure 1.1. Schematic for refraction of light at the interface between two positive indexed media. The incident wave in medium 1 is transformed at the surface into a reflected wave in medium 1 and into a refracted wave in medium 2. The incident wave is in the second quadrant and for the positive media the refracted wave is in the fourth quadrant.

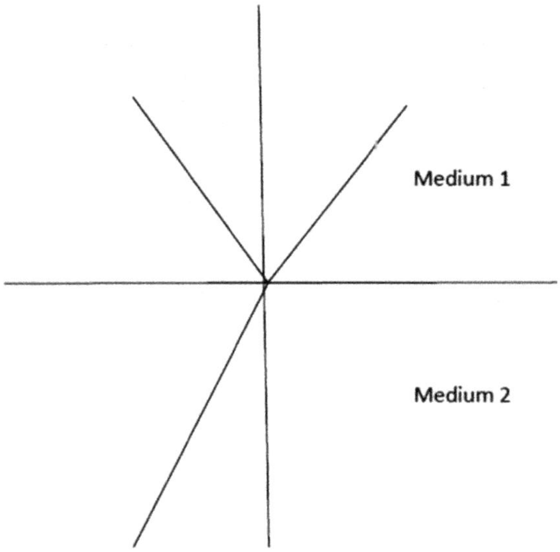

Figure 1.2. Schematic for refraction of light at the interface between positive indexed medium 1 and negative indexed medium 2. The incident wave in medium 1 is transformed at the surface into a reflected wave in medium 1 and into a refracted wave in medium 2. The incident wave is in the second quadrant and for the negative index medium the refracted wave is in the third quadrant.

related to the periodicity of the dielectric medium, have been found to mimic some of the properties of metamaterials.

The positive and negative indexed materials are alternatively referred to as right- and left-handed materials due to the different relationship of the three orthogonal vectors of the electric vector, magnetic vector, and wave vector between the two types of systems [26]. Indeed, it can be shown that waves in right-hand materials propagate electromagnetic energy parallel to the wave vector while waves in left-hand materials propagate electromagnetic energy antiparallel to the wave vector. This can have effects for example on the Cherenkov radiation in right- and left-handed materials, respectively, as well as of course for the refractive properties and the properties of antenna radiation [14, 24, 27].

1.4 Perfect lenses

One particularly interesting applications of metamaterials is in the design of so-called perfect lenses [22]. The functioning of the perfect lens is based on the new refractive property of negative index of refraction. The bending of light at the interface of a positive and negative indexed material from the second to the third quadrant allows a planar surface to form a focused image as light passes through planar surfaces. This is due to the increased bending of the light as it goes between the two media over that found between any two positive indexed media.

In the absence of negative refractive indexed materials, a curved surface is needed between two positive indexed media to focus light as it passes between the two media. This is the operating principle in the design of telescope and microscope lenses.

The design of lenses with curved surfaces limits their size. Consequently, only light passed through the aperture of the lens can be focused, and this limits the wavelength of light which will pass through the lens and become part of the image located at the focus of the lens [28]. Light must have a wavelength less than the aperture of the lens to reach the focus and, consequently, the image formed by the lens is not perfect as the loss of some wavelength components of the image decreases the resolution of the image.

With a perfect lens formed of negative refractive indexed material only planar surfaces are needed to make a focusing lens. In addition, the permittivity and permeability can be adjusted so that even the evanescent components of light from the object are reassembled at the image, giving a complete characterization in the image of the object. In principle an image with perfect resolution can be formed by imaging with an infinite slab of negative index medium. There are, however, a number of technical difficulties in the practical application of the ideas of a perfect lens. Due to the resonant nature of the materials designed to exhibit negative refractive index, the frequencies over which these properties are exhibited are limited. In addition, resonant structures tend to exhibit losses, well known from the Kramers–Kronig relations.

1.5 Periodicity

Both photonic crystals and metamaterials can be formed by the repetition of mesoscopic features on a periodic lattice. In the case of photonic crystals the interest is in the manipulation of electromagnetic radiation with wavelengths of order of the lattice constant, while metamaterials are used in the management of radiation with wavelengths much larger than the lattice constant. Metamaterials appear to be homogeneous while the important properties of the photonic crystals arise from their detailed periodic structure. For both photonic crystal and metamaterial systems formed from linear dielectric media the light propagating in the systems is electromagnetic waves satisfying linear wave equations. In metamaterials the excitations are simple plane waves and the effect of the media on the propagating waves is characterized by the effective ε and μ of the metamaterial. For the photonic crystal, on the other hand, a full account of the periodicity of the dielectric media must be given so that the excitations are more complex than simple plane waves, and a variety of methods for the calculation of the electromagnetic band structure of these materials are available. These methods are well known, having initially been developed for the treatment of the motion of electrons in the periodic ion background of semi-conductors and metals. Among those methods discussed later are the plane wave expansion method and the method of Wannier functions [2, 4, 10]. In addition, computer simulations based on finite difference time domain methods [29] and the method of moments [30] are useful in determining the system properties.

The plane wave expansion technique for photonic crystals is based on Fourier transforming the equations of motion of the systems and studying their properties in frequency–wave vector space. These methods give a good account of the properties of linear photonic crystals and are quick and easy to implement. Computer simulation methods in space and time or space and frequency can also be applied to determine the dispersion relation and field properties of the modes of the photonic crystal system. Such computer simulations are particularly important in applications treating waveguides and impurities in both photonic crystal and metamaterial systems. The treatment of waveguides and impurities is also a strength of the more analytic methods of Wannier functions [10, 31, 32]. In the method of Wannier functions, an orthogonal set of basis functions which are localized in space are generated. This set is then used to expand the modes of photonic crystals and study their properties. All of these methods will be discussed later.

1.6 Excitations in nonlinear media

While linear systems are easily handled numerically, nonlinear systems in which the dielectric properties depend on the amplitude of the excitations they support are much more challenging. In nonlinear systems, a linear combination of the wave functions of two separate solutions of the system is not a wave function solution of the system, i.e., the principle of linear superposition of modes is no longer valid. The violation of the principle of linear superposition is easily seen in the wave-like modes of nonlinear systems [7]. Due to the amplitude dependence of the dielectric, wave solutions of nonlinear systems have dispersion relations that depend on the amplitude of the wave. Consequently, increasing or decreasing the amplitude of a wave changes the frequency of its oscillation, and this change in frequency is not accounted for in a simple linear combination of plane wave solutions.

In addition, nonlinear systems exhibit new types of excitations not found in linear systems. Examples of these are solitons [9, 33–36]. Solitons occur as pulses and kinks that propagate through nonlinear optical systems. They exist due to the dependence of the dielectric properties of the systems on the amplitude of the excitations. This can be understood as follows: in a system composed solely of a linear dielectric it is possible to bind a localized electromagnetic mode to a finite region of space by appropriately changing the value of the linear dielectric in that finite region. In line with these ideas, introducing a localized electromagnetic pulse into a nonlinear dielectric system can be done in such a way that the change in the nonlinear dielectric constant caused by the changing pulse intensity supports the localized pulse. The localized pulse in turn then supports the change in the nonlinear dielectric. Such a pulse can be stationary or it can travel through the system with a constant velocity. The pulse is known as a bright soliton because of its intensity maximum. These ideas also apply to kinks. For the case of kink excitations the electric field intensity exhibits a dip rather than an intensity peak. The kink intensity shapes the dielectric response of the system which in turn supports the kink wave function. As in the case of bright solitons, kinks can also be either stationary or move through the system with constant velocities. If the intensity of the dip goes to zero the kink is known as a dark soliton, and if the intensity dip does not go to zero the kink is known as a gray soliton. In addition to the single soliton solutions there are solutions involving a number of solitons scattering from one another.

The general types of solitons we will look at are bright, dark, and gray solitons. The bright, dark, and gray solitons that are found in our photonic crystals and metamaterials are further classified by the nature of their wave functions [33, 35] into simple pulsed solitons and envelope solitons. Simple pulsed solitons occur as smooth, slowly varying pulses or a smooth, slowly varying twisting of the amplitudes of the electromagnetic fields in space. Simple pulses do not exhibit any sinusoidal modulations of the basic wave function envelopes of the excitations. Envelope solitons, on the other hand, are more complex, being composed of plane waves modulated by an envelope in the form of a pulse or kink. For these cases the solitons may be obtained in the continuum limit of the system or in the discrete lattice of the system. In discrete lattice systems they are often referred to as intrinsic localized modes or discrete breathers in the limit where they are stationary in the system. These distinctions will be discussed later.

Later the theory of soliton-like modes is discussed for both the discrete and continuum limits of photonic crystal and metamaterial systems [35]. The solutions in the discrete lattice systems are shown to be easily accessible to a variational treatment in which the wave functions are expressed as linear expansions in an appropriate set of basis states defined over the discrete lattice. Solutions of the form of bright, dark, and gray soliton-like excitations are obtained [33, 35]. In the continuum limit of the lattice systems descriptions in terms of the nonlinear Schrödinger and Klein–Gordon equations are given for photonic crystal and metamaterial systems. Both systems are found to exhibit closed form soliton-like solutions, and the properties of the bright, dark, and the gray solitons found in these systems are discussed. The relationships between the soliton-like excitations in the nonlinear Schrödinger and Klein–Gordon equations are also demonstrated.

1.7 Systems with defects and disorder

An important factor entering the study of optical components is the effects on the system arising from the introduction of disorder into the problem [37–49]. Disorder can enter the problem in the form of a mild or small renormalization of properties or with increasing disorder as a transition of the system to a whole new range of characteristic behaviors [37–40]. This change of behavior is essentially a phase transition similar to other magnetic or chemical phase transitions [10]. In the present case, however, the phase transition is observed in a plot of the system transport properties as a function of the intensity of disorder in the materials.

With the introduction into a system of mild disorder the studies of its properties can be made using analytical or computer simulation methods which treat the renormalized forms of the modal excitations in the materials. As the disorder increases a point is reached in the increasing degree of disorder of the system at which a fundamental change occurs in the modal excitations of the materials [10, 37–40]. In this transition the functional forms of the wave function are transformed to new types. An example of this is the introduction of disorder into a homogeneous medium which supports plane wave modal excitations.

With weak disorder the excitations are renormalized plane waves which extend throughout the material. At a point of increased disorder in the medium, however, a transition is made so that the modes in the material become localized modes.

Specifically, the localized modes are restricted to a finite region of space in the materials. This modal transition changes the transport properties of the material, and in electrical conductors is observed as a metal–insulator transition known as the Anderson transition [10, 37]. Similar transitions from conductive to non-conductive behaviors with the presence of increasing disorder are observed in optical and acoustic materials.

Related to random disorder, site impurities and lines or clusters of impurities can also be purposely introduced into systems in order to form various types of resonant structures and waveguides. These may have various technological applications and may exist in engineered materials in addition to design imperfections. They form an important topic of technological interest and are introduced in the context of the applications of group theory techniques in their study

Consequently, as a final topic in the book an introduction is given to the study of disorder in optical media [34–49]. This will include some basic studies of single site and multiple site impurities in photonic crystals and metamaterials. For this treatment, some elements of group theory are introduced into the discussion of cluster impurities. This is followed by an introduction of Anderson localization. First a treatment of localization is given in the context of conductivity in electronic systems [10, 37]. This is followed by the occurrence of the localization transition in the treatments of transport properties of optical systems. As a final point, we conclude with the discussion of weak Anderson localization in the diffuse reflection of light from randomly rough planar and periodic surfaces [44, 49], and in various effects in the presence of nonlinearity.

The order of the book is to first begin with a treatment of the dielectric properties of materials. The properties and models of linear photonic crystals and metamaterials are next discussed along with the basic methods used to compute their properties. This is followed by discussions of nonlinear photonic crystal and metamaterial models and their theoretical treatments. Various soliton modes and discrete breathers found in these systems are presented along with some final comments on computer simulation studies. To conclude, a treatment of sites, clusters, and waveguide impurities are given, followed by a general discussion of Anderson localization.

References

[1] Joannopoulos J D, Vilenueve P R and Fan S 1995 *Photonic Crystals* (Princeton, NJ: Princeton University Press)

[2] Sakoda K 2001 *Optical Properties of Photonic Crystals* (Berlin: Springer)

[3] Favennec P N 2005 *Photonic Crystals: Towards Nanoscale Photonic Devices* (Berlin: Springer)

[4] McGurn A R 2002 *Survey of Semiconductor Physics* ed W Boer (New York: Wiley) ch 13

[5] Waser R (ed) 2005 *Nanoelectronics and Information Technology: Advanced Electronic Materials and Novel Devices* 2nd edn (Weinheim: Wiley)

[6] Boyd R W 2003 *Nonlinear Optics* 2nd edn (Amsterdam: Academic)

[7] Mills D L 1998 *Nonlinear Optics* (Berlin: Springer)

[8] Banerjee P P 2004 *Nonlinear Optics* (New York: Dekker)

[9] Kivshar Y S and Agrawal G P 2003 *Optical Solitons* (Amsterdam: Academic)

[10] Marder M P 2010 *Condensed Matter Physics.* 2nd edn (Hoboken, NJ: Wiley)

[11] Noda S, Fujita M and Asano T 2007 Spontaneous-emission control by photonic crystals and nanocavities *Nat. Photonics* **1** 449

[12] Agrawal G P 2012 *Fiber-Optics Communication Systems* 4th edn (New York: Wiley)

[13] Yeh P 2005 *Optical Waves in Layered Media* (New York: Wiley)

[14] Engheta N and Ziolkowski R W (ed) 2006 *Metamaterials: Physics and Engineering Explorations* (New York: Wiley-IEEE)

[15] Giri P, Choudhary K, Sen Gupta A, Bandyopadhyay A K and McGurn A R 2011 Klein–Gordon equation approach to nonlinear split-ring resonator based metamaterials: one-dimensional systems *Phys. Rev.* B **84** 155429

[16] Eleftheriou M, Lazarides N and Tsironis G P 2008 Magnetoinductive breathers in metamaterials *Phys. Rev.* E **77** 036608

[17] Eleftheriades G V 2009 EM transmission-line metamaterials *Mater. Today* **12** 30

[18] Kourakis I, Lazarides N and Tsironis G P 2007 Self-focusing and envelope pulse generation in nonlinear magnetic metamaterials *Phys. Rev.* E **75** 067601

[19] Lapine M, Gorkunov M and Ringhofer K H 2003 Nonlinearity of a metamaterial arising from diode inserts into resonant conductive elements *Phys. Rev.* E **67** 065601

[20] Shadrivov I V, Morrison S K and Kivshar Y 2006 Tunable split-ring resonators for nonlinear negative-index metamaterials *Opt. Express* **14** 9344

[21] Tamayama Y, Nakanishi T and Kitano M 2013 Observation of modulation instability in a nonlinear magnetoinductive waveguide *Phys. Rev.* B **87** 195123

[22] Pendry J B 2000 Negative refraction makes a perfect lens *Phys. Rev. Lett.* **85** 3966

[23] Agranovich V M and Gartstein Y N 2006 Spatial dispersion and negative refraction of light *Phys.–Usp.* **49** 1029

[24] Agranovich V M 2006 Hybrid organic–inorganic nanostructures and light–matter interaction *Problems of Condensed Matter Physics* ed A L Ivanov and S G Tikhodeev (Oxford: Clarendon) ch 2

[25] Ramakrishna S A 2005 Physics of negative index materials *Rep. Prog. Phys.* **68** 449

[26] Veselago V G 1968 The electrodynamics of substances with simultaneously negative values of ε and μ *Sov. Phys.–Usp.* **10** 509

[27] Cai W and Shalaev V 2001 *Optical Metamaterials: Fundamental and Applications* (New York: Springer)

[28] Smith W J 2007 *Modern Optical Engineering* 4th edn (New York: McGraw-Hill)

[29] Taflove A 1995 *Computational Electrodynamics: The Finite-Difference Time-Domain Method* (Boston, MA: Artech House)

[30] El Misilmani H M, Kabalan K Y, Abou-Shahine M Y and Al-Husseini M 2015 A method of moment approach in solving boundary value problems *J. Electromagn. Anal. Appl.* **7** 61

[31] McGurn A R 2005 Impurity mode techniques applied to the study of light sources *J. Phys.* **38** 2338

[32] McGurn A R 1996 Green's-function theory for row and periodic defect arrays in photonic band structures *Phys. Rev.* B **53** 7059

[33] Dauxois T and Peyrard M 2006 *Physics of Solitons* (Cambridge: Cambridge University Press)

[34] Sievers A J and Page J B 1995 *Dynamical Properties of Solids* ed G K Horton and A A Maradudin (Amsterdam: North-Holland)

[35] McGurn A R 2013 Barrier transmission map of one-dimensional nonlinear split-ring-resonator-based metamaterials: bright, dark, and gray soliton resonances *Phys. Rev.* B **88** 155110
[36] McGurn A R 2008 Transmission through nonlinear barriers *Phys. Rev.* B **77** 115105
[37] Lee P A and Ramakrishnan T V 1985 Disordered electron systems *Rev. Mod. Phys.* **57** 287
[38] Segev M, Silverberg Y and Christodoulides D 2013 Anderson localization of light *Nat. Photonics* **7** 197
[39] McGurn A 2020 *Introduction to Photonic and Phononic Crystals and Metamaterials* (San Rafael, CA: Morgan & Claypool)
[40] Shing P (ed) 1990 *Scattering and Localization of Classical Waves in Random Media* (Singapore: World Scientific)
[41] McGurn A R 2018 *Nanophotonics* (Cham: Springer)
[42] Cai W and Shalaev V 2009 *Optical Metamaterials* (Berlin: Springer)
[43] McGurn A R, Maradudin A A, Christensen K and Mueller F 1993 Anderson localization in one-dimensional randomly disordered optical systems that are periodic on average *Phys. Rev.* B **47** 13120
[44] McGurn A R 1990 Enhanced retroreflectance effects in the reflection of light from randomly rough surfaces *Surf. Sci. Rep.* **10** 357
[45] McGurn A R and Maradudin A A 1994 Localization of light in linear and non-linear random layered dielectric systems that are periodic on average in the proceeding of Third Int. Conf. on Electrical Transport and Optical Properties of Inhomogeneous Media *Physica* A **207** 435
[46] Devillard P and Souillard B 1986 Polynomial decaying transmission for the nonlinear Schrodinger equation in a random medium *J. Stat. Phys.* **43** 423
[47] McGurn A R, Sheng P and Maradudin 1992 Strong localization of light in two-dimensional disordered dielectric media *Opt. Commun.* **91** 175
[48] de Raedt H, Lagemdijk A and de Vries P 1989 Transverse localization of light *Phys. Rev. Lett.* **62** 47
[49] McGurn A R, Leskova T A and Agranovich V M 1991 Weak localization effects in the generation of second harmonics of light at a randomly rough vacuum-metal grating *Phys. Rev.* B **44** 11441

Arthur R McGurn

Chapter 2

A simple model of nonlinear dielectrics

A simple model for the electric susceptibility of dielectric materials is based on simulating the atomic electrons as bound in classical oscillators with the atomic polarizability determined from the oscillator response to an externally applied electric field [1]. For a linear optical system the equation of motion of the model is

$$m[\ddot{x} + \gamma\dot{x} + \omega_0^2 x] = -eE \tag{2.1}$$

where x is the electron displacement, m the electron mass, γ the phenomenological decay rate of the oscillator accounting for dissipative losses to the crystal, and ω_0 the angular frequency of the electron motion. The externally applied electric field causing the polarizing response is of the form $E = \tilde{E}\mathrm{e}^{-\mathrm{i}\omega t} + cc$ (here cc indicates the complex conjugate of the preceding term) and enters on the right as a forcing term. From the solution of (2.1) the polarization of a system of N atoms, simplistically defined by $P = -eNx$, is then of the form of a Lorentzian response

$$P = \frac{e^2N}{\omega_0^2 - \omega^2 - \mathrm{i}\gamma\omega}\tilde{E}\mathrm{e}^{-\mathrm{i}\omega t} + cc. \tag{2.2}$$

The result in (2.2) provides a basic indicator of the qualitative polarization properties of a dielectric material, exhibiting a resonance at the ω_0 oscillator frequency which is limited by the oscillator decay rate. In a perfected model, self-consistent field effects, often accounted for by simulation methods or crudely by relations such as the Clausius–Mossotti equation [1], need to be included as well as a quantum mechanical treatment given to the system. Considerations of the possible wave vector dependence in the polarization arising from many-body effects due to interactions between the N atoms may also be important. The general form in (2.2), however, is often successful in developing a semi-quantitative understanding of the response of many systems. Its form has been used to fit experimental data, treating

doi:10.1088/978-0-7503-3579-9ch2

the system response as a sum over a sequence of dielectric resonances each modeled by (2.2).

The model can also be extended to consider nonlinear optical systems [2, 3]. For these the nonlinearity enters as weak perturbation terms so that (2.1) becomes

$$m[\ddot{x} + \gamma\dot{x} + \omega_0^2(x + \alpha x^2 + \beta x^3)] = -eE \tag{2.3}$$

where the quadratic and cubic perturbations are sources of a variety of well-known nonlinear optical effects. In low orders of perturbation theory the αx^2 term is responsible for a response of the system at frequency 2ω known as second harmonic generation and a response of the system at zero frequency. It also eventually leads to a renormalization of the frequency of the ω_0 mode of the atom. Similarly, at lower orders the βx^3 term causes the generation of a third harmonic response at frequency 3ω and a Kerr renormalization of the electric susceptibility at the ω mode. In a higher order the cubic term also gives a renormalization of the frequency of the ω_0 mode of the isolated atom. These effects, as they arise from each of the two perturbation terms, are separately considered in the following. We will begin by discussing the effects of nonlinearity on the oscillator frequencies. This provides a nice introduction to the specialties of nonlinear perturbation theory. Following this the nonlinear polarization responses of the system are handled.

To begin, consider an isolated atom in the absence of an external field, treating the effects of the weak nonlinearity on the oscillator frequency. The renormalization of the electron frequency mode is obtained from (2.3) using nonlinear perturbation theory. First consider the case of cubic nonlinearity in a system without dielectric losses, taking $\alpha = 0$, $\gamma = 0$, and $E = 0$. The resulting equation is of the form

$$m[\ddot{x} + \omega_0^2(x + \beta x^3)] = 0 \tag{2.4}$$

and a small dimensionless parameter characterizing the strength of the nonlinearity of the equation in terms of β and the atomic radius, a_0, is given by $\varepsilon = \beta a_0^2$. With this parameterization of the nonlinear interaction, a solution of the electronic displacement is written in terms of a series expansion in ε of the form

$$x = x_0 + \varepsilon x_1 + \varepsilon^2 x_2 + \cdots \tag{2.5}$$

where the coefficients $\{x_n\}$ to be determined have units of length. In the case of weak perturbations, the series in (2.5) can often be limited to a few of its lowest order terms. In treating (2.4) and (2.5), however, it is important to note that for the most effective development of the series in (2.5) special care must be taken in its handling. A closer look at this is given in the following.

To most effectively develop the perturbation series of nonlinear systems, a general result from the study of the nonlinear perturbation theory must be applied [4]. Specifically, in nonlinear systems it is necessary to treat the time in a special way so as to avoid divergences in the perturbation series. Rather than treating the time t as the sole variable describing the time dependence of the system, the time dependence of nonlinear systems should be described by an infinite set of time variables scaled in powers of the small parameter, ε. These time variables are defined by the forms

$$t_0 = t,\ t_1 = \varepsilon t, \ldots\ldots,\ t_m = \varepsilon^m t\ldots\ldots\ldots\ldots \tag{2.6}$$

Consequently, in (2.4) and (2.5) we should take

$$x_n = x_n(t_0, t_1, t_2, \ldots.) \tag{2.7}$$

rather than the $x_n = x_n(t)$ form used in the perturbation theory treatment of linear systems. In terms of the additional variables the time derivatives in (2.4) become

$$\begin{aligned}\frac{\mathrm{d}}{\mathrm{d}t} &= \frac{\partial}{\partial t_0} + \varepsilon\frac{\partial}{\partial t_1} + \cdots\cdot+\varepsilon^n\frac{\partial}{\partial t_n} + \cdots. \\ &= D_0 + \varepsilon D_1 + \cdots\cdots\cdots + \varepsilon^n D_n + \cdots..\end{aligned} \tag{2.8}$$

where $D_i = \frac{\partial}{\partial t_i}$.

Using (2.7) and (2.8) in (2.4) a series equation in powers of ε is generated. Each order ε^n in this series gives a differential equation for x_n involving a function of the x_m where $m < n$. The order ε^0 term in the series equation gives

$$D_0^2\ddot{x}_0 + \omega_0^2 x_0 = 0 \tag{2.9}$$

so that $x_0 = \tilde{x}_0 + \tilde{x}_0^*$ with $\tilde{x}_0 = \chi_0 \mathrm{e}^{-\mathrm{i}\omega_0 t_0}$. At the next order in the series equation, the ε^1 order term gives x_1 obtained from

$$D_0^2 x_1 + \omega_0^2 x_1 = -\omega_0^2\frac{x_0^3}{a_0^2} - 2D_0 D_1 x_0. \tag{2.10}$$

Note that for a solution of (2.10) the forcing terms on the right-hand side of (2.10) cannot contain a time dependence of the form $\mathrm{e}^{\pm\mathrm{i}\omega_0 t_0}$. Such terms would drive the oscillator form on the left-hand side at its resonant frequency causing a singularity in x_1. This divergence can be removed by requiring the dominant $\mathrm{e}^{\pm\mathrm{i}\omega_0 t_0}$ terms on the right-hand side of (2.10) to be set to zero, i.e., this requires that

$$2D_0 D_1 \tilde{x}_0 + 3\omega_0^2\left|\frac{\tilde{x}_0}{a_0}\right|^2 \tilde{x}_0 = 0. \tag{2.11}$$

To satisfy (2.11) write $\tilde{x}_0\ (t_0, t_1, t_2, \ldots.) = \chi_0\ (t_1, t_2, ..)\ \mathrm{e}^{-\mathrm{i}\omega_0 t_0} = \chi_{01}\ (t_2, ..)\mathrm{e}^{-\mathrm{i}(\omega_0 t_0 + \omega_1 t_1)}$, where the time dependence of $\tilde{x}_0$ on t_0 and t_1 is written out explicitly. The D_1 factor in (2.11) then introduces an additional unknown ω_1 into (2.11) which is chosen to satisfy (2.11), i.e., to set the terms in $\mathrm{e}^{\pm\mathrm{i}\omega_0 t_0}$ on the right-hand side of (2.10) to zero. Choosing ω_1 to accomplish this, the resonant driving terms on the right-hand side of (2.10) are then removed while higher harmonic $\mathrm{e}^{\pm 3\mathrm{i}\omega_0 t_0}$ and zero frequency terms remain. A result is the determination of ω_1. This represents the renormalization of the ω_0 frequency arising from the nonlinearity. Solving (2.11) gives $\tilde{x}_0 = A\mathrm{e}^{-\mathrm{i}(\omega_0+\beta\omega_1)t_0}$ where $\omega_1 = \frac{3}{2}\omega_0\ |A|^2$ and A is the $t_0 = 0$ amplitude of $\tilde{x}_0$. From the resulting general solution of (2.9) and (2.10) a well behaved expansion in powers of the nonlinearity is obtained of the form

$$x = \tilde{x}_0 + \varepsilon\tilde{x}_1 + cc = A\mathrm{e}^{-\mathrm{i}(\omega_0+\beta\omega_1)t} + \frac{\beta}{8}A^3\mathrm{e}^{-\mathrm{i}3\omega_0 t} + cc. \tag{2.12}$$

Going to higher powers of the series equation, generates an equation for general x_n of the form

$$D_0^2 x_n + \omega_0^2 x_n = f(x_0, x_1, \ldots\ldots, x_{n-1}). \tag{2.13}$$

Here $f(x_0, x_1, \ldots \ldots, x_{n-1})$ is a nonlinear function consisting of a sum of time-dependent terms at different frequencies. Again the focus is on the subset of these terms with an $e^{\pm i\omega_0 t_0}$ time-dependence. These terms would resonantly drive the oscillator form on the left-hand side of (2.13). The resonant driving terms in $e^{\pm i\omega_0 t_0}$ on the right-hand side of (2.13) are handled in a similar manner to those in (2.10) and (2.11), yielding a correction ω_n to the original oscillator frequency ω_0.

The development of the perturbation series in the time variables in (2.6) is found to introduce at each order of perturbation an extra degree of freedom (e.g., ω_1 in the case of (2.11)) into the mathematics. This allows for the removal of the $e^{\pm i\omega_0 t_0}$ dependent so-called secular terms which resonantly drive the oscillator at ω_0 and arise from powers of the lower order perturbation terms. These powers of the lower order perturbation terms can only arise from the nonlinear terms of the system and are absent in the standard perturbation theory of linear systems. This is why linear perturbation theory only requires the single time variable for its development.

A similar analysis is made for the isolated atom with a nonlinearity of the form αx^2. The atomic equation of motion is

$$m\,[\ddot{x} + \omega_0^2(x + \alpha x^2)] = 0 \tag{2.14}$$

and the perturbation series defined in (2.5) and (2.6) are now expansions in $\varepsilon = \alpha a_0$ where a_0 is an atomic radius. Proceeding as in the case of βx^3 gives to order ε

$$\begin{aligned} x &= \tilde{x}_0 + \varepsilon \tilde{x}_1 + cc \\ &= A e^{-i(\omega_0 + \alpha^2\omega_2)t} + \alpha\left[\frac{1}{3}A^2 e^{-i2\omega_0 t} - |A|^2\right] + \frac{\alpha^2}{12}A^3 e^{-i3\omega_0 t} + cc \end{aligned} \tag{2.15}$$

where

$$\omega_2 = -\frac{5}{3}\omega_0\,|A|^2. \tag{2.16}$$

In both systems the nonlinearity causes the single frequency modes of the linear system to evolve into excitations composed of multi-frequency components, and the original oscillator frequency is renormalized by the amplitude of the electron motion. These effects carry over to the treatment of the nonlinear oscillators driven by an outside field and are observed in their dielectric properties.

Consider the nonlinear polarization of (2.3) for the case in which $\beta x^3 = 0$ and the dissipation term, γx, is not small compared to the nonlinearity. Applying (2.5) and (2.6), and expanding in the small parameter $\varepsilon = \alpha a_0$, the zero order term is defined by

$$D_0^2 x_0 + \gamma D_0 x_0 + \omega_0^2 x_0 = -\frac{eE}{m} \tag{2.17}$$

with solution

$$x_0 = \frac{-e\tilde{E}/m}{\omega_0^2 - \omega^2 - \mathrm{i}\gamma\omega} e^{-\mathrm{i}\omega t} + cc. \tag{2.18}$$

To first order in ε

$$D_0^2 x_1 + \gamma D_0 x_1 + \omega_0^2 x_1 = -\omega_0^2 \frac{x_0^2}{a_0} - 2D_0 D_1 x_0 - \gamma D_1 x_0 \tag{2.19}$$

with solution

$$\varepsilon x_1 = -\frac{\alpha(e\tilde{E}/m)^2}{\left(\omega_0^2 - \omega^2\right)^2 + \gamma^2\omega^2} - \frac{\omega_0^2}{\omega_0^2 - 4\omega^2 - \mathrm{i}2\gamma\omega} \frac{\alpha(e\tilde{E}/m)^2}{[\omega_0^2 - \omega^2 - \mathrm{i}\gamma\omega]^2} e^{-\mathrm{i}2\omega t} + cc. \tag{2.20}$$

To this order $x = x_0 + \varepsilon x_1$, and the polarization, $P = -Nex = -Ne(x_0 + \varepsilon x_1)$, is seen to contain terms at the original frequency of the electromagnetic field, twice the original frequency, and constant in time. In addition, the polarization develops a dependence on the square of the applied field from the higher order perturbations.

The calculation of the nonlinear polarization proceeds similarly for the case in which $\alpha x^2 = 0$ in (2.3) and the dissipation term, γx, is not small compared to the nonlinearity. Applying (2.5) and (2.6) for an expansion in $\varepsilon = \beta a_0^3$ gives to order ε^0

$$D_0^2 x_0 + \gamma D_0 x_0 + \omega_0^2 x_0 = -\frac{eE}{m} \tag{2.21}$$

with a solution

$$x_0 = \frac{-e\tilde{E}/m}{\omega_0^2 - \omega^2 - \mathrm{i}\gamma\omega} e^{-\mathrm{i}\omega t} + cc. \tag{2.22}$$

The correction to order ε^1 is given by

$$D_0^2 x_1 + \gamma D_0 x_1 + \omega_0^2 x_1 = -\omega_0^2 \frac{x_0^3}{a_0^2} - 2D_0 D_1 x_0 - \gamma D_1 x_0 \tag{2.23}$$

with a solution

$$\begin{aligned} \varepsilon x_1 = &\frac{3\omega_0^2}{\omega_0^2 - \omega^2 - \mathrm{i}\gamma\omega} \frac{|e\tilde{E}/m|^2}{\left(\omega_0^2 - \omega^2\right)^2 + \gamma^2\omega^2} \frac{\beta(e\tilde{E}/m)}{[\omega_0^2 - \omega^2 - \mathrm{i}\gamma\omega]} e^{-\mathrm{i}\omega t} \\ &+ \frac{\omega_0^2}{\omega_0^2 - 9\omega^2 - \mathrm{i}3\gamma\omega} \frac{\beta(e\tilde{E}/m)^3}{[\omega_0^2 - \omega^2 - \mathrm{i}\gamma\omega]^3} e^{-\mathrm{i}3\omega t} + cc. \end{aligned} \tag{2.24}$$

Again from these results $x = x_0 + \varepsilon x_1$, and the polarization, $P = -Nex = -Ne(x_0 + \varepsilon x_1)$, contains terms at the original frequency of the electromagnetic field and three times the original frequency. The higher order perturbation terms now introduce an optical nonlinearity in the form of a cubic dependence of the polarization on the applied electric field. For both the quadratic and cubic nonlinearities the behavior of the system driven at the resonance frequency is determined to a great extent by the dissipative term in (2.3).

An interesting case is that in which the dissipative term in (2.3) is small relative to the nonlinearity and the interactions with the external fields are of the order of the nonlinear term, i.e., $\gamma = \varepsilon^n \Gamma$ for $n > 1$ and $E = \varepsilon\Sigma = \tilde{E}e^{-i\omega_0 t} + cc$. Considering an external field at resonance with the oscillator for $\alpha = 0$ so only the cubic nonlinearity is present, yields an expansion of (2.3) in $\varepsilon = \beta a_0^2$. To zero order

$$D_0^2 \ddot{x}_0 + \omega_0^2 x_0 = 0 \tag{2.25}$$

with $x_0 = \tilde{x}_0 + \tilde{x}_0^*$ where $\tilde{x}_0 = \chi_0 e^{-i\omega_0 t_0}$, and to order ε^1

$$D_0^2 x_1 + \omega_0^2 x_1 = -\omega_0^2 \frac{x_0^3 + 3\,|x_0|^2\, x_0}{a_0^2} - 2D_0 D_1 x_0 - \frac{e\Sigma}{m}. \tag{2.26}$$

Requiring that there are no forcing terms of frequency ω_0 on the right-hand side of (2.26) gives a solution to order ε of the form

$$x = Ae^{-i(\omega_0 + \beta\omega_1)t} + \frac{\beta A^3}{8} e^{-i3\omega_0 t} + cc \tag{2.27}$$

where $\omega_1 = \frac{3}{2}\,|A|^2\,\omega_0 + \frac{e\tilde{E}}{2m\omega_0\beta A}$. For the polarization, $P = -Nex$, to this order, the frequency of the polarization resonance is observed to shift with both the amplitude of the electron motion and that of the external electric field. This is different from the behavior observed in the off resonance Kerr dielectric.

The considerations in ((2.25)–(2.27)) can also be applied to an external electric field at resonance with an oscillator for which $\alpha \neq 0$ but $\beta = 0$, i.e., only the quadratic nonlinearity is present. Upon expanding (2.3) in $\varepsilon = \alpha a_0$ for $E = \varepsilon^2\Sigma = \tilde{E}e^{-i\omega_0 t} + cc$, a solution to order ε^2 of the form

$$x = Ae^{-i(\omega_0 + \alpha^2\omega_2)t} + \alpha\left[\frac{1}{3}A^2 e^{-i2\omega_0 t} - |A|^2\right] + \frac{\alpha^2 A^3}{12} e^{-i3\omega_0 t} + cc \tag{2.28}$$

where $\omega_2 = -\frac{5}{3}\,|A|^2\,\omega_0 + \frac{e\tilde{E}}{2m\omega_0\alpha^2 A}$ is obtained. For the polarization, $P = -Nex$, the frequency dependence of the polarization resonance is observed to shift with both the amplitude of the electron motion and that of the external electric field. This is similar to that found in (2.27) for the cubic nonlinearity.

To conclude this simplified treatment of the nonlinear properties of optical media, we note two important results that will be used in later discussions of photonic crystals and metamaterials. First, the nonlinear properties that will be of most interest to us later are those arising from the cubic terms of the nonlinear oscillator in (2.3). Specifically, from (2.24) it is seen that in a system described by the cubic term, a polarization response is generated at the frequency of the applied field and the strength of the electric susceptibility of the response depends on the absolute square of the field amplitude. This field-dependent response is the origin of the Kerr nonlinear term in optics, and it causes the permittivity and dielectric constant of a Kerr nonlinear optical medium to depend on the absolute square of the field amplitude of an applied electromagnetic wave. A second important result comes from the quadratic term in (2.3). From (2.20) it is seen that this term introduces a polarization response at double the frequency of the applied field. It is responsible for the generation of second harmonic radiation from applied radiation at the fundamental frequency. Second harmonic generation will be treated at the end of the book. The other polarization terms, though important, are not further discussed. One final point to note is that the nonlinear potential for the Kerr effect is symmetric in the electron displacement while that for second harmonic generation is asymmetric. This is a general property of these responses and the reader is referred elsewhere for a more detailed treatment [5].

References

[1] Jackson J D 1999 *Classical Electrodynamics* 3rd edn (New York: Wiley)
[2] Mills D L 1998 *Nonlinear Optics* (Berlin: Springer)
[3] Banerjee P P 2004 *Nonlinear Optics* (New York: Dekker)
[4] Nayfeh A H 1981 *Introduction to Perturbation Techniques* (New York: Wiley)
[5] Boyd R W 2003 *Nonlinear Optics* 2nd edn (Amsterdam: Academic)

IOP Publishing

Introduction to Nonlinear Optics of Photonic Crystals and Metamaterials (Second Edition)

Arthur R McGurn

Chapter 3

Basic properties of photonic crystals

In this chapter the basic properties of linear media photonic crystals are discussed along with some of the theoretical methods for determining these properties. The general features arising in the physics of excitations in a periodic medium are developed with an introduction to some of the standard terms that are used to describe these properties, i.e., direct and reciprocal lattices, general forms of the wave functions, Brillouin zones, etc. Following this the theoretical approaches of the plane wave method, Wannier functions, finite-difference time-domain methods, and methods of moments are outlined as techniques in understanding the excitations in photonic crystals. The chapter concludes with some discussion of periodic surfaces, fiber optic waveguides, and fiber Bragg gratings. First we start with a comparison of the typical features of the excitations in free space and in periodic dielectric media, with a development of the theoretical structures to understand the differences in the excitations in these two types of systems.

3.0 Comparison of free space and periodic media solutions

The dispersion relation of an electromagnetic wave is strongly influenced by the translational symmetry of the medium in which the wave propagates [1–4]. In free space an electromagnetic wave of frequency, ω, moving in the x–y plane with an electric field, $E\ (x,\ y)$, polarized parallel to the z-axis is described by the two-dimensional Helmholtz equation

$$\left[\frac{\partial^2}{\partial x^2} + \frac{\partial^2}{\partial y^2} + \frac{\omega^2}{c^2}\right] E\ (x,\ y)\ = 0. \tag{3.1}$$

Due to the translational symmetry of (3.1) over the entire x–y plane, the energy densities of its frequency modes exhibit complete translational symmetry in the x–y plane. Consequently, the modes are plane waves with $E(x,\ y) \propto e^{i(\vec{k} \cdot \vec{r} - \omega t)}$ for $\vec{k}$ and

doi:10.1088/978-0-7503-3579-9ch3

$\vec{r}$ vectors in the x–y plane, representing uniform energy densities defined over space and proportional to $| E |^2$.

For the case in which the wave moves in a two-dimensionally periodic dielectric medium, the Helmholtz equation (in Gaussian units) for motion parallel to the plane of the periodicity becomes [1–4]

$$\left[\frac{\partial^2}{\partial x^2} + \frac{\partial^2}{\partial y^2} + n(x, y)^2\frac{\omega^2}{c^2}\right]E\,(x, y) = 0 \tag{3.2}$$

where $n(x, y) = \sqrt{\varepsilon(x, y)\mu(x, y)}$ is a periodic function of both x and y, and the electric field is polarized parallel to the z-axis. (As an example, figure 3.1(a) presents a schematic in the x–y plane of a two-dimensional periodic rectangular direct lattice. The periodic functions in x and y defined on this lattice would exhibit the basic periodicity shown in the figure.) As with (3.1), the energy densities of the modal solutions of (3.2) reflect the periodicity of the dielectric function, $n(x, y)$. In the particular limit $n(x, y)\rightarrow 1$, complete translational symmetry is restored, and the wave functions of (3.2) become plane wave solutions of (3.1) each with a uniform energy density. For an arbitrary periodic dielectric in the x–y plane, a general form satisfying the requirements of both the periodicity properties of the energy density and the limiting $n(x, y) \rightarrow 1$ plane wave nature of the solutions is given by

$$E(x, y) = b_{\vec{k}}(x, y)\mathrm{e}^{-\mathrm{i}\omega t} = \mathrm{e}^{\mathrm{i}\left[\vec{k}\,\bullet\,\vec{r}-\omega t\right]}u_{\vec{k}}(x, y), \tag{3.3}$$

where $\vec{k}$ and $\vec{r} = x\hat{i} + y\hat{j}$ are vectors in the x–y plane, and $u_{\vec{k}}(x, y)$, for each $\vec{k}$, is a periodic function with the periodicity of $n(x, y)$.

The eigenmodes of (3.2) are then of the form given in (3.3), and the modes of different frequencies are orthogonal. A complication arises, however, in formulating the orthogonality properties of the modes defined by (3.2). This is due to the association of the factor $n(x, y)^2$ with the eigenvalue in (3.2). In the Strum–Liouville theory [5] of differential equation eigenvalue problems $n(x, y)^2$ enters the

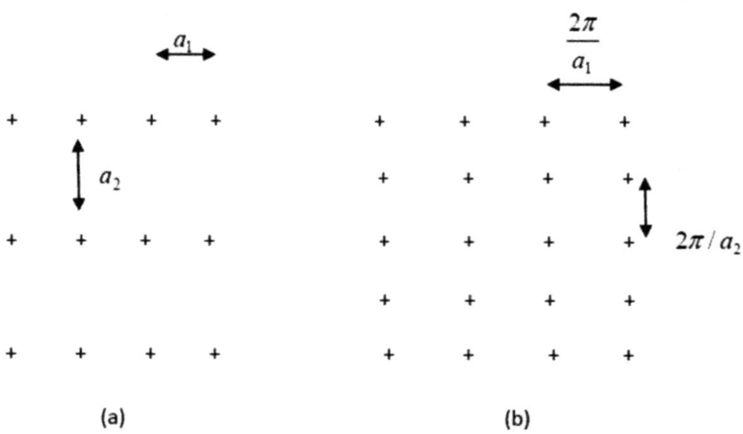

Figure 3.1. Direct and reciprocal rectangular lattices: (a) direct and (b) reciprocal lattices. Here the direct lattice is defined in the space of spatial position through which the wave moves, and the reciprocal lattice is defined in the space of possible wave vectors representing the wave functions.

orthonormality relations for the modes as a weight function. Hence for a typical photonic crystal with a position independent $\mu(x, y) = \mu$ so that $n(x, y)^2 = \varepsilon(x, y)\mu$, the different eigenmodes $b_{i,\vec{k}}(\vec{r})$ and $b_{j,\vec{k}'}(\vec{r})$, where i and j label eigenmodes of different frequency for the same $\vec{k}$, have the orthonormality property

$$\int \varepsilon(\vec{r}) b*_{i,\vec{k}}(\vec{r}) b_{j,\vec{k}'}(\vec{r}) \mathrm{d}\vec{r} = \delta_{i,j}\delta_{\vec{k},\vec{k}'}. \tag{3.4}$$

An additional property of the solutions of (3.2) is that they are characterized by $\vec{k}$ in a bounded region of phase space and are periodic functions of $\vec{k}$ [6] over all $\vec{k}$*-space*. If $\vec{a}_1$ and $\vec{a}_2$ are the two smallest linearly independent vectors in the x–y plane that translate $n(x, y)$ into itself, then applying the translation $\vec{T} = m_1\vec{a}_1 + m_2\vec{a}_2$ for m_1 and m_2 integers to $n(x, y)$ leaves it invariant but changes (3.3) to

$$E_{\vec{T}}(x, y) = \mathrm{e}^{\mathrm{i}\left[\vec{k}\bullet\vec{r}-\omega t\right]} u_{\vec{k}}(x, y) \mathrm{e}^{\mathrm{i}\vec{k}\bullet\vec{T}}. \tag{3.5}$$

An important special case of (3.5) is that for which $\vec{k}$ has the general form $\vec{k} = \vec{G} = n_1\vec{g}_1 + n_2\vec{g}_2$ for n_1 and n_2 integers with the $\vec{g}_i$ satisfying $\vec{g}_i \bullet \vec{a}_j = 2\pi\delta_{i,j}$. Under these conditions $\vec{k} \bullet \vec{T} = 2\pi N$ for an integer N, and $E(x, y)$ is periodic in the x–y plane with the periodicity of $n(x, y)$. For more general $\vec{k}$ it is found from (3.3) that the solutions of (3.2) can be rewritten as

$$E(x, y) = \mathrm{e}^{\mathrm{i}\left[\left(\vec{k}-\vec{G}\right)\bullet\vec{r}-\omega t\right]} u_{\vec{k}}(x, y) \mathrm{e}^{\mathrm{i}\vec{G}\bullet\vec{r}} = \mathrm{e}^{\mathrm{i}\left[\left(\vec{k}-\vec{G}\right)\bullet\vec{r}-\omega t\right]} \tilde{u}_{\vec{k}-\vec{G}}(x, y) \tag{3.6}$$

where both $\mathrm{e}^{\mathrm{i}\vec{G}\bullet\vec{r}}$ and $\tilde{u}_{\vec{k}-\vec{G}}(x, y)$ have the periodicity of $n(x, y)$. Equation (3.6) is a demonstration that the wave functions at $\vec{k}$-$\vec{G}$ are in a one-to-one correspondence with those at $\vec{k}$. This periodicity in phase space also applies to the frequency dispersion relation of the system in (3.2), which is shown by substituting (3.3) and the far right-hand side of (3.6) separately into (3.2). Doing this gives

$$\nabla^2 u_{\vec{k}} + 2\mathrm{i}\vec{k} \bullet \nabla u_{\vec{k}} - k^2 u_{\vec{k}} + n^2\frac{\omega^2(\vec{k})}{c^2} u_{\vec{k}} = 0 \tag{3.7}$$

and

$$\nabla^2 \tilde{u}_{\vec{k}-\vec{G}} + 2\mathrm{i}\left(\vec{k} - \vec{G}\right) \bullet \nabla\tilde{u}_{\vec{k}-\vec{G}} - k^2\tilde{u}_{\vec{k}-\vec{G}} + n^2\frac{\omega^2\left(\vec{k} - \vec{G}\right)}{c^2}\tilde{u}_{\vec{k}-\vec{G}} = 0 \tag{3.8}$$

for all $\vec{G}$. The eigenvalue spectra of (3.7) and (3.8) have the same solutions but are shifted in wave vector space by $\vec{G}$. Consequently, the dispersion relation of the electric field eigenmodes is periodic in the lattice in wave vector space with the periodicity defined by the translation vectors $\{\vec{G}\}$. The periodic lattice defined by the $\{\vec{G}\}$ is known as the reciprocal lattice and is illustrated in figure 3.1(b) for the rectangular direct lattice of figure 3.1(a).

For a periodic dielectric we see that the general form of the wave function is a plane wave multiplied by a function which is periodic in the direct lattice, e.g., the space in figure 3.1(a). The dispersion relation of the wave function eigenmodes depends on the wave vector of the plane wave parts of their wave functions and is a periodic function in reciprocal space, e.g., the space in figure 3.1(b). These general results are useful in applying the various methods of solutions which we now discuss.

We begin by looking at plane wave methods. The plane wave method is a natural approach to use in developing an understanding of the physics of periodic systems.

3.1 Plane wave method

An often used approach for finding the eigenvalues and eigenvectors of (3.2) is the plane wave expansion method [2–4]. In this method (3.3) is written in the form

$$E(x, y) = \mathrm{e}^{\mathrm{i}\left[\vec{k}\bullet\vec{r}-\omega t\right]}\sum_{\vec{G}} A(\vec{k}, \vec{G})\mathrm{e}^{\mathrm{i}\vec{G}\bullet\vec{r}} \tag{3.9}$$

where $\vec{G} = m_1\vec{g}_1 + m_2\vec{g}_2$ for $\{m_1, m_2\}$ integers and $\{\vec{g}_1, \vec{g}_2\}$ are defined below (3.5). Similarly, for the periodic dielectric

$$n(x, y) = \sum_{\vec{G}} n(\vec{G})\mathrm{e}^{\mathrm{i}\vec{G}\bullet\vec{r}}. \tag{3.10}$$

Substituting (3.9) and (3.10) into (3.2) and using the orthogonality properties of the plane waves gives a matrix eigenvalue problem for $\{A(\vec{k}, \vec{G})\}$ and $\omega^2(\vec{k})$ determined at fixed $\vec{k}$.

As an illustration, in [7] a solution of the eigenvalue problem was given for a square lattice of cylindrical dielectric rods. The system parameters were chosen for a comparison with a system that was previously studied experimentally [7]. In figure 3.2 these results are presented for the frequency band structure of electromagnetic waves

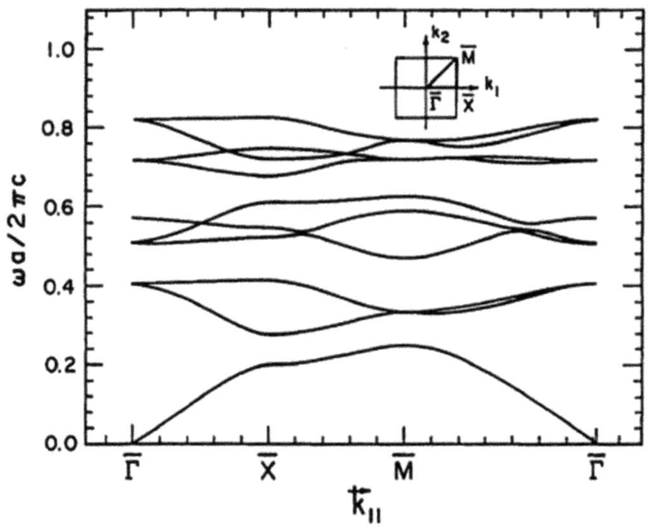

Figure 3.2. Band structure of frequency versus wave vector in the first Brillouin zone. Directions in k-space in the first Brillouin zone are shown in the inset. Reproduced from [7] by permission of The Optical Society of America.

along directions of the smallest unit cell in phase space containing a complete set of solutions. The cylindrical rods are of dielectric constant 9 in a background of air or a vacuum with a filling fraction of 0.4488. The modes propagate in the plane perpendicular to the cylinder axes with the electric field polarized parallel to the cylinder axes, and the dimensionless results are presented for a lattice constant of the square lattice denoted, a. Since the dispersion relation of the electromagnetic modes is periodic in the two-dimensional wave vector space, the modes' dispersion is often presented in the smallest square cell of the reciprocal lattice, known as the first Brillouin zone. This smallest cell has no repetition of the dispersion, only presenting unique solutions of the system. The inset in the figure shows the first Brillouin zone and denotes a set of lines in k-space along which the frequency solutions are plotted in the figure.

3.2 Wannier function method

A method that is useful for the study of impurities in photonic crystals is based on the techniques of Wannier functions [6, 8]. In this method the Block plane wave states of (3.3) are used to create an alternative basis set of localized orthogonal functions known as Wannier functions. These functions are particularly suited to discuss the properties of fields localized about impurities and within waveguides. The periodic dielectric constant of the photonic crystal can be considered as composed as a repetition of a single basic dielectric unit throughout space. Each such repeated unit is related to the other repetitions by a translation vector, $\vec{T} = m_1\vec{a}_1 + m_2\vec{a}_2$. The Wannier functions are chosen to describe the local behavior of the Block wave functions of the system within a basic repeat unit of the periodic dielectric so that the Block functions are expressed as a sum over the crystal lattice of localized Wannier functions multiplied by a phase. This separates the local behavior of the Block waves from their translational symmetry within the periodic lattice of the entire crystal. In the following, the development of the Wannier functions is made by analogy with the relationship between free space plane wave solutions of (3.1) and a delta function pulse of localized free space electromagnetic radiation fields.

For the free space system in (3.1) the electric field modes are of the form

$$E = \frac{1}{2\pi}\mathrm{e}^{\mathrm{i}\left(\vec{k}\,\bullet\,\vec{r}-\omega t\right)}. \tag{3.11}$$

Although these modes are of uniformly extended intensity throughout space they can be combined to express a highly localized electric pulse. In particular, at $t = 0$ a localized pulse of radiation described by

$$2\pi\delta(\vec{r}) = \frac{1}{2\pi}\int \mathrm{d}\vec{k}\,\mathrm{e}^{\mathrm{i}\vec{k}\,\bullet\,\vec{r}} \tag{3.12}$$

is made from a uniformly weighted sum of the plane wave modes of (3.11). Adding a phase to the integrand in (3.12) so that $\mathrm{e}^{\mathrm{i}\vec{k}\,\bullet\,\vec{r}}$ is replace by $\mathrm{e}^{\mathrm{i}\vec{k}\,\bullet\,(\vec{r}-\vec{R})}$, the position of the pulse can also be shifted to any location in the x–y plane. Consequently, a basis formed of localized pulses is generated from the original basis set of extended plane waves.

Similar sets of relationships to those in (3.11) and (3.12) can be developed in the Block wave basis of a periodic lattice. In this development the counterparts of the free space plane waves and delta functions are, respectively, the extended Block waves and the localized Wannier functions. These are now discussed.

Following the treatment in (3.11) and (3.12), a localized pulse of radiation in the periodic system described by (3.2) is made from the extended Block modes in (3.3). To facilitate this, the Block wave notation in (3.3) is extended so that

$$E(x, y) = b_{n,\vec{k}}(\vec{r})\mathrm{e}^{-\mathrm{i}\omega t} = \mathrm{e}^{\mathrm{i}\left[\vec{k}\,\bullet\,\vec{r}-\omega t\right]}u_{n,\vec{k}}(x, y) \tag{3.13}$$

where the additional subscript n is a band index labeling the different frequency eigenmodes at fixed $\vec{k}$. (For example, in figure 3.2 we see that there are a number of different band solutions for a fixed $\vec{k}$.) The band index arises upon the solution of the eigenvalue problem and the realization that there are many different frequency modes corresponding to the same $\vec{k}$. In this notation, by analogy with (3.12), we define from (3.3) the localized functions of position

$$a_{n,(m_1,\, m_2)}(\vec{r}) = \frac{1}{\sqrt{N}}\sum_{\vec{k}}\mathrm{e}^{-\mathrm{i}\vec{k}\,\bullet\,\vec{R}}b_{n,\vec{k}}(\vec{r}) \tag{3.14}$$

where $\vec{R} = m_1\vec{a}_1 + m_2\vec{a}_2$ and N is the number of lattice sites. The function defined in (3.14) acts as one of a set of basis functions localized in the vicinity of $\vec{r} = m_1\vec{a}_1 + m_2\vec{a}_2$ for the modes of band index n. (For an example, note that in the case of a uniform $b_{n,\vec{k}}(\vec{r}) \propto \mathrm{e}^{\mathrm{i}\vec{k}\,\bullet\,\vec{r}}$, the associated $a_{n,(m_1,\, m_2)}(\vec{r})$ would be a pulse which is localized about $(\vec{r}-\vec{R})\approx 0$.) As with (3.12) the localization is a result of the constructive and destructive interference of the plane wave Block forms.

In addition, for any translation vector of the periodic lattice, $\vec{T} = m_1\vec{a}_1 + m_2\vec{a}_2$, the function defined by

$$a_{n,(m_1,\, m_2)}(\vec{r}) = \frac{1}{\sqrt{N}}\sum_{\vec{k}}\mathrm{e}^{\mathrm{i}\vec{k}\,\bullet\,(\vec{r}-\vec{T})}u_{n,\vec{k}}(\vec{r}) = a_{n,(0,0)}(\vec{r} - \vec{T}) \tag{3.15}$$

is $a_{n,(0,0)}(\vec{r})$ shifted by a lattice translation. This follows from the translational invariance of $u_{\vec{k}}(x, y)$ within the periodic lattice. In the discussions of the properties of the system in (3.2), the set of all functions $\{a_{n,(m_1,m_2)}(\vec{r})\}$ can be shown to form a localized orthogonal basis known as the Wannier functions, and from (3.15) the members of the set are all related to one another by translation vectors of the crystal lattice. The orthogonality relations of the Wannier functions are now demonstrated.

The orthogonality of the set $\{a_{n,(m_1,\, m_2)}(\vec{r})\}$ follows directly from that of the Block waves in (3.13). To see this consider the integral

$$\begin{aligned}\int \varepsilon(\vec{r})a^{*}_{n,(m_1,\, m_2)}(\vec{r})a_{n',(m'_1,\, m'_2)}(\vec{r})\mathrm{d}\vec{r} &= \int \varepsilon(\vec{r})a^{*}_{n,(0,0)}(\vec{r} - \vec{R})a_{n',(0,0)}(\vec{r} - \vec{R}')\mathrm{d}\vec{r} \\ &= \frac{1}{N}\sum_{\vec{k},\vec{k}'}\mathrm{e}^{\mathrm{i}\left[\vec{k}\,\bullet\,\vec{R}-\vec{k}'\,\bullet\,\vec{R}'\right]}\int \varepsilon(\vec{r})\left(u^{*}_{n,\vec{k}}(\vec{r})\mathrm{e}^{-\mathrm{i}\vec{k}\,\bullet\,\vec{r}}\right)\left(u_{n',\vec{k}'}(\vec{r})\mathrm{e}^{\mathrm{i}\vec{k}'\,\bullet\,\vec{r}}\right)\mathrm{d}\vec{r} \\ &= \delta_{(m_1,\, m_2),(m'_i,\, m'_2)}\delta_{n,n'}\end{aligned} \tag{3.16}$$

where $\vec{R} = m_1 a_1 \hat{i} + m_2 a_2 \hat{j}$, $\vec{R}' = m'_1 a_1 \hat{i} + m'_2 a_2 \hat{j}$, and the translation and orthonormality properties of the Block waves are used. Equation (3.16) is then a statement of the orthonormality of the Wannier functions, providing a basis for the expansion of the system properties in states localized about the individual sites in the direct crystal lattice. A general expansion in Wannier functions, based on (3.16), can then be written as

$$\psi(\vec{r}) = \frac{1}{\sqrt{N}} \sum_{j=1}^{N} F(\vec{R}_j) a_{n,(0,0)}(\vec{r} - R_j) \tag{3.17}$$

where $\vec{R}_j = m_1 a_1 \hat{i} + m_2 a_2 \hat{j}$ and $F(\vec{R}_j)$ are expansion coefficients. As an example of (3.17), the Block waves of the system are written in terms of the Wannier basis functions, giving

$$b_{n,\vec{k}}(\vec{r}) = \frac{1}{\sqrt{N}} \sum_{j=1}^{N} e^{i\vec{k} \bullet \vec{R}_j} a_{n,(0,0)}(\vec{r} - R_j). \tag{3.18}$$

In (3.18), the Wannier functions $a_{n,(0,0)}(\vec{r})$ are seen to provide the localized variation about the lattice sites but are multiplied at each site by a phase so that the overall translational symmetry of the Block wave solutions is maintained. Equations (3.17) and (3.18) are particularly interesting results for the treatment of impurity problems. These are now considered.

Consider the case of an impurity problem based on (3.2) and defined by [8]

$$[H_0 + V(\vec{r})]\psi(\vec{r}) = \varepsilon(\vec{r})\frac{\omega^2}{c^2}\psi(\vec{r}). \tag{3.19}$$

Here $H_0 = -\frac{\partial^2}{\partial x^2} - \frac{\partial^2}{\partial y^2}$, we have taken $n^2(\vec{r}) = \varepsilon(\vec{r})$ for the pure photonic crystal, and $V(\vec{r}) = -\delta\varepsilon(\vec{r})\frac{\omega^2}{c^2}$ is the position-dependent impurity potential arising from a change $\delta\varepsilon(\vec{r})$ of the dielectric constant from that of the pure photonic crystal. Substituting (3.17) into (3.19), multiplying by $a^*_{n,(0,0)}(\vec{r} - \vec{R}_i)$, and integrating over $\vec{r}$ changes the differential equation problem into a set of difference equations for the envelope function $F(\vec{R}_j)$

$$\sum_{j'} \left[h_0(\vec{R}_{j'}) F(\vec{R}_j - \vec{R}_{j'}) + V_{j,j'} F(\vec{R}_{j'}) \right] = \frac{\omega^2}{c^2} F(\vec{R}_j) \tag{3.20}$$

where

$$h_0(\vec{R}_j - \vec{R}_{j'}) = \int d\vec{r} a^*_{n,(0,0)}(\vec{r} - \vec{R}_j) H_0 a_{n,(0,0)}(\vec{r} - \vec{R}_{j'}) \tag{3.21}$$

and

$$V_{j,j'} = \int d\vec{r} a^*_{n,(0,0)}(\vec{r} - \vec{R}_j) V(\vec{r}) a_{n,(0,0)}(\vec{r} - \vec{R}_{j'}). \tag{3.22}$$

Equation (3.20) yields a solution for the envelope function $F(\vec{R}_j)$ of the Wannier function expansion and offers an advantage over the full impurity wave function solutions of (3.19). The rapid changes in the full wave functions of the impurity modes are accounted for in (3.17) by the localized Wannier functions and the space dependence of the envelope function tends to be smooth and slowly varying in space. With regard to this notice that, in the case that $V(\vec{r}) = 0$, (3.20) reduces to the tight binding Hamiltonian of the photonic crystal in the absence of impurities. The envelope functions for this case are plane waves whereas the complete wave functions of (3.19) are rapidly varying in space due to the localized nature of the Wannier functions (i.e., see also (3.18)). In the presence of an impurity, we shall see that the localized nature of the Wannier functions along with the localized nature of the impurity potential means that $V_{j,j'}$ quickly goes to zero as the j, j' site separation increases. This leads to a simplified problem for the envelope function involving linear algebra.

A difficulty with (3.20) is that it is a difference equation which connects the eigenmode solutions at many different sites. It will now be shown that under some general conditions a continuum limit of (3.20) can be made which reduces the problem of determining the envelope function defined in (3.17) to a standard type of Schrödinger differential equation problem [6, 8]. The Schrödinger equation is now for the envelope function rather than for the detailed rapidly varying wave function of the complete impurity problem.

It is first noted that some simplification can be made to (3.20) by considering the first term on the left-hand side of the equation which is of the general form $\sum_{j'} h_0(\vec{R}_{j'})F(\vec{r} - \vec{R}_{j'})$. The term can be rewritten in a Taylor series expansion in r as

$$\sum_{j'} h_0\left(\vec{R}_{j'}\right)F\left(\vec{r} - \vec{R}_{j'}\right) = \sum_{j'} \mathrm{e}^{-\vec{R}_{j'} \bullet \nabla} h_0\left(\vec{R}_{j'}\right)F(\vec{r}) \tag{3.23}$$

where ∇ operates on (x, y). This expansion is now used to develop the continuum limit Schrödinger equation for $F(\vec{r})$. In order to understand this limit, notice that when $V(r) = 0$, (3.20) is solved by a $F(\vec{R}_j) \propto \mathrm{e}^{-\mathrm{i}\vec{k} \bullet \vec{R}_j}$ plane wave with an eigenvalue solution

$$\sum_{j'} h_0\left(\vec{R}_{j'}\right)\mathrm{e}^{-\mathrm{i}\vec{k} \bullet \vec{R}_{j'}} = \frac{\omega_0^2(\vec{k})}{c^2} \tag{3.24}$$

where $\omega_0(\vec{k})$ is a frequency in the pure system. In proceeding to the continuum limit it is assumed that $F(\vec{R}_j) \propto \mathrm{e}^{-\mathrm{i}\vec{k} \bullet \vec{R}_j}$ becomes a general function of space $F(\vec{r}) \propto \mathrm{e}^{-\mathrm{i}\vec{k} \bullet \vec{r}}$ and that the introduction of $V(\vec{r})$ only modifies the F with slowly varying envelopes. Under these considerations the operator on the right-hand side of (3.23) satisfies an eigenvalue problem similar to the result in (3.24), i.e.,

$$\sum_{j'} h_0\left(\vec{R}_{j'}\right)\mathrm{e}^{-\vec{R}_{j'} \bullet \nabla} F(\vec{r}) = \frac{\omega_0^2(\vec{k})}{c^2} F(\vec{r}) \tag{3.25a}$$

where $F(\vec{r}) \propto \mathrm{e}^{-\mathrm{i}\vec{k} \bullet \vec{r}}$ is the plane wave eigenmode.

Taken one step further, the wave vectors in k-space can be identified with spatial differential operators by using the quantum mechanical $\vec{k} \leftrightarrow -i\nabla$ correspondence. With this correspondence (3.25a) becomes

$$\sum_{j'} h_0(\vec{R}_{j'}) e^{-\nabla \bullet \vec{R}_{j'}} F(\vec{r}) = \frac{\omega_0^2(-i\nabla)}{c^2} F(\vec{r}). \tag{3.25b}$$

This equation is valid for the system in the absence of the impurity potential and is a good approximation when the amplitude of $F(\vec{r})$ is of a slow spatial variation. From (3.23) it then follows in the continuum limit that

$$\sum_{j'} h_0(\vec{R}_{j'}) F(\vec{r} - \vec{R}_{j'}) = \frac{\omega_0^2(-i\nabla)}{c^2} F(\vec{r}) \tag{3.26}$$

which with (3.20) yields the impurity equation

$$\left[\frac{\omega_0^2(-i\nabla)}{c^2} + V(\vec{r})\right] F(\vec{r}) = \frac{\omega^2}{c^2} F(\vec{r}). \tag{3.27}$$

Here it is assumed that $V(r)$, as an approximation for $V_{j,j'}$, is a short ranged potential compared to the spatial variation of $F(\vec{r})$, i.e., the approximation $V_{j,j'} \approx V(\vec{R}_j)\delta_{j,j'}$ involving the continuum limit of the impurity potential is made.

In the impurity systems focused on nano-optics, the impurity modes of interest are bound states with energies in the stop bands of the photonic crystals. For these systems the frequency term, $\frac{\omega_0^2(\vec{k})}{c^2}$, in (3.27) is at an extreme upper or lower bound of the pass band of the pure system. Consequently, it can often be approximated as a parabolic band by $\frac{\omega_0^2(\vec{k})}{c^2} = E_0 + \frac{\hbar^2}{2m} k^2$ for the purpose of determining the stop band bound states. (Note: E_0 and $\frac{\hbar^2}{2m}$ are parameters of the parabolic fit to the band extrema in a notation chosen to make the impurity problem in (3.27) resemble the impurity problem for electron levels in semiconductors. For the photonic crystal systems, however, the units of these two parameters are different from those of the electron problem.) With this notation $\frac{\hbar^2}{2m} > 0$ for an impurity state below the bottom of a pass band and $\frac{\hbar^2}{2m} < 0$ for an impurity state above the top of a pass band. Using the parabolic approximation in (3.25) the impurity equation (3.27) becomes

$$\left[-\frac{\hbar^2}{2m}\nabla^2 + V(\vec{r}) + E_0 - \frac{\omega^2}{c^2}\right] F(\vec{r}) = 0. \tag{3.28}$$

A common choice of impurity potential is $V(\vec{r}) = -\delta\varepsilon(\vec{r})\frac{\omega^2}{c^2}$ where $\delta\varepsilon(\vec{r})$ is the change in the dielectric of the photonic crystal due to the impurity [8]. Equation (3.28) is then written in the form of a Schrödinger equation impurity problem:

$$\left[-\frac{\hbar^2}{2m}\nabla^2 + E_0 - \frac{\omega^2}{c^2}\right]F(\vec{r}) = \delta\varepsilon(\vec{r})\frac{\omega^2}{c^2}F(\vec{r}). \tag{3.29}$$

The solution of the inhomogeneous differential equation in (3.29) gives the envelope function, $F(\vec{r})$, of the impurity modes at frequency ω in the stop band of the pure photonic crystal system. The problem is easily treated using standard Green function methods [9–11]. In this approach, the Green function for (3.29), $G(\vec{r},\omega)$, is obtained as a solution of

$$\left[-\frac{\hbar^2}{2m}\nabla^2 + E_0 - \frac{\omega^2}{c^2}\right]G(\vec{r}, \omega) = \delta(\vec{r}). \tag{3.30}$$

For bound state boundary conditions it is given by

$$G(\vec{r}, \omega) = \frac{1}{2\pi}\frac{2m}{\hbar^2}K_0\left(\left[\frac{2m}{\hbar^2}\left(E_0 - \frac{\omega^2}{c^2}\right)\right]^{1/2} r\right) \tag{3.31}$$

where $K_0(x)$ is a modified Bessel function of the second kind. From (3.31) the solution of (3.29) is then formally obtained as

$$F(\vec{r}) = \frac{\omega^2}{c^2}\int d\vec{r}\,'G(\vec{r} - \vec{r}\,', \omega)\delta\varepsilon(\vec{r}\,')F(\vec{r}\,'). \tag{3.32}$$

When considered for a frequency $\frac{\omega^2}{c^2}$ in the stop band of the photonic crystal, this equation can often be converted into an integral equation eigenvalue problem. In the eigenvalue treatment the envelope of the impurity mode, $F(\vec{r})$, is the eigenvector and the eigenvalues are the change in the dielectric constant of the impurity from the feature it replaces in the pure photonic crystal, i.e., the eigenvalues are related to $\delta\varepsilon(\vec{r})$. As an illustration (3.32) is now discussed for a single impurity.

Consider a single site impurity that is formed by replacing a cylinder of the photonic crystal and let the radii of the cylinders of the photonic crystal and impurity be R. If $\delta\varepsilon_{00}$ is the change in the dielectric constant of the impurity dielectric constant from that of the cylinder of the photonic crystal it is replacing, then (3.32) for the modes bound to the impurity becomes [8]

$$F(\vec{s}) = e\int_S d\vec{s}\,'K_0(|\,\vec{s} - \vec{s}\,'|)F(\vec{s}\,'). \tag{3.33}$$

Here $\vec{s} = [(2m/\hbar^2)(E_0 - (\omega^2/c^2))]^{1/2}\vec{r}$, the eigenvalue is of the form $e = \frac{1}{2\pi}\frac{(\omega^2/c^2)\delta\varepsilon_{00}}{E_0 - (\omega^2/c^2)}$, and the region of integration S is a circle of radius $s_0 = [(2m/\hbar^2)(E_0 - (\omega^2/c^2))]^{1/2}R$ centered about the origin of the photonic crystal lattice. Equation (3.33) is solved for e as a function of E_0, $\hbar^2/2m$, and ω^2/c^2. Once the eigenvalue e and eigenvector $F(s)$ are determined, the relation

$$\delta\varepsilon_{00} = 2\pi\left(\frac{E_0}{\omega^2/c^2} - 1\right)e \tag{3.34}$$

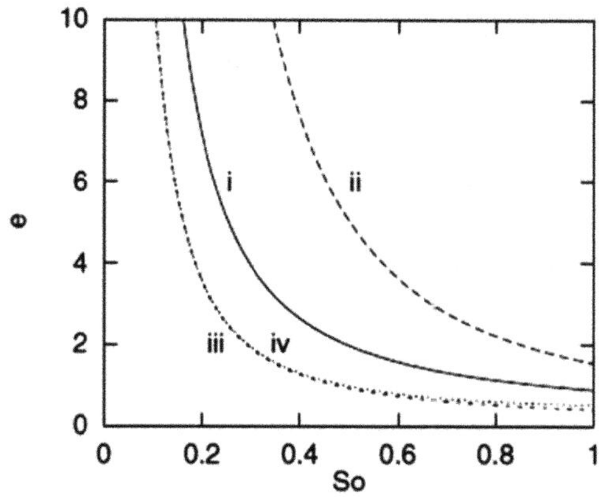

Figure 3.3. Plot of e versus s_0. Reproduced from [8] © IOP Publishing.

gives the values of $\delta\varepsilon_{00}$ which will support an impurity mode of frequency ω within the stop band of the photonic crystal.

In figure 3.3 results for e plotted versus s_0 from the numerical solution of (3.33) are presented [8]. The curves labeled *i* and *ii* are, respectively, the lowest and next to lowest eigenvalues of a series of eigenvalues. The figure is taken from [8] which contains additional discussion of the other labeled curves that are not related to the discussion presented here. The figure shows that for a fixed s_0 there are a series of different e and, consequently, different $\delta\varepsilon_{00}$ which support eigenvector envelope solutions, $F(\vec{R})$.

3.3 Computer simulation methods

Computer simulation methods fall into three general categories: finite-difference time-domain methods [12], the method of moments [13, 14], and finite element methods [15]. All of these are based on discretizing the four Maxwell's equations into sets of difference equations which are solved by machine over a finite region of space and/or time. In the following the focus is on the propagation and scattering of electromagnetic waves in the absence of net electric and magnetic charges for problems involving regions containing dielectric and magnetic materials. Much of the recent work on computer simulations is performed using the standard international form of the Maxwell equations and we adopt these for the simulation discussion presented here.

For the discretization it is convenient to take Maxwell's equations in the form [12]:

$$\frac{\partial \vec{B}}{\partial t} = -\nabla \times \vec{E} - \vec{J}_{\mathrm{m}} \qquad (3.35a)$$

$$\frac{\partial \vec{D}}{\partial t} = \nabla \times \vec{H} - \vec{J}_{\mathrm{e}} \qquad (3.35b)$$

$$\nabla \bullet \vec{D} = 0 \qquad (3.35c)$$

$$\nabla \bullet \vec{B} = 0 \tag{3.35d}$$

with constituent relations

$$\vec{B} = \mu\vec{H} \tag{3.36a}$$

$$\vec{D} = \varepsilon\vec{E} \tag{3.36b}$$

$$\vec{J}_{\text{m}} = \sigma^{\text{m}}\vec{H} \tag{3.36c}$$

$$\vec{J}_{\text{e}} = \sigma^{\text{e}}\vec{E} \quad . \tag{3.36d}$$

Here the possibility of both a magnetic current, $\vec{J}_{\text{m}}$, as well as an electric current, $\vec{J}_{\text{e}}$, is included. These currents are related to the magnetic and electric fields through the magnetic conductivity, σ^{m}, and the electrical conductivity, σ^{e}. The inclusion of the possibility of both electric and magnetic currents is found to be useful in designing simulations meant to study scattered waves. Later, in the discussions of boundary conditions it will be seen that by properly arranging the electrical and magnetic conductivities of an outer boundary layer, a perfectly absorbing region which does not reflect radiation incident upon it can be arranged. This arrangement is known as the perfectly matching layer (PML) type of absorbing boundary condition (ABC) [12, 16] and is useful at the outer edges of the necessarily bounded spatial region of a computer simulation. It eliminates radiation that arrives at the outer spatial edges of a simulation from the interior of the simulation without reflecting components of the radiation back into the scattering region. Any such reflected components of radiation would give rise to spurious results in the scattering generated by the simulation which is meant to approximate the scattering in an infinite system.

3.3.1 Finite-difference time-domain method

The finite-difference time-domain method is considered for a two-dimensional system in which the dielectric and magnetic properties of the system do not depend on the z-coordinates and the waves propagate in the plane perpendicular to the z-axis. For this system (3.35) and (3.36) become [12]

$$\frac{\partial H_x}{\partial t} = -\frac{1}{\mu}\left(\frac{\partial E_z}{\partial y} + \sigma^{\text{m}}H_x\right) \tag{3.37a}$$

$$\frac{\partial H_y}{\partial t} = \frac{1}{\mu}\left(\frac{\partial E_z}{\partial x} - \sigma^{\text{m}}H_y\right) \tag{3.37b}$$

$$\frac{\partial H_z}{\partial t} = \frac{1}{\mu}\left(\frac{\partial E_x}{\partial y} - \frac{\partial E_y}{\partial x} - \sigma^{\text{m}}H_z\right) \tag{3.37c}$$

$$\frac{\partial E_x}{\partial t} = \frac{1}{\varepsilon}\left(\frac{\partial H_z}{\partial y} - \sigma^{\text{e}}E_x\right) \tag{3.37d}$$

$$\frac{\partial E_y}{\partial t} = -\frac{1}{\varepsilon}\left(\frac{\partial H_z}{\partial x} + \sigma^{\mathrm{e}} E_y\right) \tag{3.37e}$$

$$\frac{\partial E_z}{\partial t} = \frac{1}{\varepsilon}\left(\frac{\partial H_y}{\partial x} - \frac{\partial H_x}{\partial y} - \sigma^{\mathrm{e}} E_z\right). \tag{3.37f}$$

These equations can in turn be decoupled into equations for transverse magnetic (TM) modes which have magnetic fields perpendicular to the z-axis and transverse electric (TE) modes having electric fields perpendicular to the z-axis. For the TM modes

$$\frac{\partial H_x}{\partial t} = -\frac{1}{\mu}\left(\frac{\partial E_z}{\partial y} + \sigma^{\mathrm{m}} H_x\right) \tag{3.38a}$$

$$\frac{\partial H_y}{\partial t} = \frac{1}{\mu}\left(\frac{\partial E_z}{\partial x} - \sigma^{\mathrm{m}} H_y\right) \tag{3.38b}$$

$$\frac{\partial E_z}{\partial t} = \frac{1}{\varepsilon}\left(\frac{\partial H_y}{\partial x} - \frac{\partial H_x}{\partial y} - \sigma^{\mathrm{e}} E_z\right) \tag{3.38c}$$

are treated, and for the TE modes

$$\frac{\partial E_x}{\partial t} = \frac{1}{\varepsilon}\left(\frac{\partial H_z}{\partial y} - \sigma^{\mathrm{e}} E_x\right) \tag{3.39a}$$

$$\frac{\partial E_y}{\partial t} = -\frac{1}{\varepsilon}\left(\frac{\partial H_z}{\partial x} + \sigma^{\mathrm{e}} E_y\right) \tag{3.39b}$$

$$\frac{\partial H_z}{\partial t} = \frac{1}{\mu}\left(\frac{\partial E_x}{\partial y} - \frac{\partial E_y}{\partial x} - \sigma^{\mathrm{m}} H_z\right) \tag{3.39c}$$

are treated. In the following the difference equations for the TM modes are obtained and discussed. The TE mode results are obtained in a similar manner and are left for the reader to work out.

The set of difference equations for the TM evolution in space and time are obtained by discretizing the differential equations on a space–time lattice [12]. It is good to begin by focusing on the discretization of (3.38c). The space–time lattice has discrete coordinates for the electric field that are of the form $(i\,\Delta\,x, j\,\Delta\,y, n\,\Delta\,t)$ for i, j, n integers. As a simplification, these coordinates may be abbreviated by just listing the integers of the lattice sites, i.e., (i,j,n). Following this notation the electric field $E_z(x, y, t)$ in (3.38) is written as

$$E_{z\,i,j}^{\;n} = E_z(i\Delta x, j\Delta y, n\Delta t) \tag{3.40}$$

at each of its discrete space–time coordinates (i,j,n). Using the conventional definition of the derivative, the time derivative of the electric field is then given by

$$\frac{\partial E_z}{\partial t}\Big|_{x=i\Delta x, y=j\Delta y, t=(n+\frac{1}{2})\Delta t} \approx \frac{E_{zi,j}^{\,n+1} - E_{zi,j}^{\,n}}{\Delta t}. \tag{3.41a}$$

Notice, however, that the time derivative is obtained at the time coordinate $t = (n + \frac{1}{2})\Delta t$. From (3.38c) a consequence of this is that, while the planes of constant time for the discretized E field are at $t = n\,\Delta\, t$, the planes of constant time for discretized H are at $t = (n + \frac{1}{2})\Delta t$. The electric and magnetic fields are seen to alternate in their time updates as the simulation proceeds forward in time.

The same reasoning is next applied to the space derivatives in (3.38c). By extending to the space domain the consideration and notation of half-integer lattice points, we define

$$\begin{aligned}
&\frac{\partial H_y}{\partial x}\Big|_{x=i\Delta x, y=j\Delta y, t=(n+\frac{1}{2})\Delta t} \\
&\approx \frac{H_y((i+\frac{1}{2})\Delta x, j\Delta y, (n+\frac{1}{2})\Delta t) - H_y((i-\frac{1}{2})\Delta x, j\Delta y, (n+\frac{1}{2})\Delta t)}{\Delta x} \\
&= \frac{H_{y_{i+\frac{1}{2},j}}^{\,n+\frac{1}{2}} - H_{y_{i-\frac{1}{2},j}}^{\,n+\frac{1}{2}}}{\Delta x}
\end{aligned} \tag{3.41b}$$

and similarly

$$\frac{\partial H_x}{\partial y}\Big|_{x=i\Delta x, y=j\Delta y, t=(n+\frac{1}{2})\Delta t} \approx \frac{H_{x_{i,j+\frac{1}{2}}}^{\,n+\frac{1}{2}} - H_{x_{i,j-\frac{1}{2}}}^{\,n+\frac{1}{2}}}{\Delta y}. \tag{3.41c}$$

These two forms give the derivatives at $(x,y) = (i\,\Delta\, x, j\,\Delta\, y)$ needed on the right-hand side of (3.38c) but writes these derivatives in terms of the magnetic fields at the sites $(x, y) = ((i + \frac{1}{2})\Delta x, j\Delta y)$ and $(x, y) = (i\Delta x, (j + \frac{1}{2})\Delta y)$. Again from a consideration of (3.38c) it is seen to be most natural to discretize H_x and H_y in the following forms

$$H_{x_{i,j+\frac{1}{2}}}^{\,n+\frac{1}{2}} = H_x\left(i\Delta x, \left(j + \frac{1}{2}\right)\Delta y, \left(n + \frac{1}{2}\right)\Delta t\right) \tag{3.42a}$$

and

$$H_{y_{i+\frac{1}{2},j}}^{\,n+\frac{1}{2}} = H_x\left(\left(i + \frac{1}{2}\right)\Delta x, j\Delta y, \left(n + \frac{1}{2}\right)\Delta t\right). \tag{3.42b}$$

These represent discretizations of the magnetic fields on space–time lattices that are shifted relative to those defined by the set of integer triple coordinates used to discretize the electric field. Observing this lattice convention, the time derivatives of the magnetic fields in (3.38) take the forms

$$\frac{\partial H_x}{\partial t}\Big|_{x=i\Delta x, y=(j+\frac{1}{2})\Delta y, t=n\Delta t} \approx \frac{H_{x\,i,j+\frac{1}{2}}^{\,n+\frac{1}{2}} - H_{x\,i,j+\frac{1}{2}}^{\,n-\frac{1}{2}}}{\Delta t} \tag{3.43a}$$

$$\frac{\partial H_y}{\partial t}\Big|_{x=(i+\frac{1}{2})\Delta x, y=j\Delta y, t=n\Delta t} \approx \frac{H_{y\,i+\frac{1}{2},j}^{\,n+\frac{1}{2}} - H_{y\,i+\frac{1}{2},j}^{\,n-\frac{1}{2}}}{\Delta t}. \tag{3.43b}$$

In addition, in order to assure the consistency of the various discretizations of the fields, the last term on the right in (3.38c) is written in the form

$$\frac{E_{zi,j}^{n+1} + E_{zi,j}^{n}}{2} \tag{3.44a}$$

where the continuity of the electric field is used. This is reasonable in the absence of free charges in the system. Similarly, in the last term on the right of (3.38a)

$$\frac{H_{x\,i,j+\frac{1}{2}}^{\,n+\frac{1}{2}} + H_{x\,i,j+\frac{1}{2}}^{\,n-\frac{1}{2}}}{2} \tag{3.44b}$$

and the last term on the right of (3.38b)

$$\frac{H_{y\,i+\frac{1}{2},j}^{\,n+\frac{1}{2}} + H_{y\,i+\frac{1}{2},j}^{\,n-\frac{1}{2}}}{2}. \tag{3.44c}$$

Indeed, using this discretization procedure in the remaining TM equations in (3.38) provides a discretization of the entire set of equations in a consistent and satisfactory manner.

Substituting (3.40) through (3.44) into (3.38) and performing a little algebra gives the time advancement difference equations for the magnetic and electric fields of the TM modes. For the magnetic fields

$$H_{x\,i,j+\frac{1}{2}}^{\,n+\frac{1}{2}} = \frac{1}{1+\dfrac{\Delta t}{2}\dfrac{\sigma^{\mathrm{m}}_{i,j+\frac{1}{2}}}{\mu_{i,j+\frac{1}{2}}}}\left[\left(1-\frac{\Delta t}{2}\frac{\sigma^{\mathrm{m}}_{i,j+\frac{1}{2}}}{\mu_{i,j+\frac{1}{2}}}\right)H_{x\,i,j+\frac{1}{2}}^{\,n-\frac{1}{2}} + \frac{\Delta t}{\Delta y}\frac{1}{\mu_{i,j+\frac{1}{2}}}\left(E_{zi,j}^{n} - E_{zi,j+1}^{n}\right)\right] \tag{3.45a}$$

$$= A_0(i,j)H_{x\,i,j+\frac{1}{2}}^{\,n-\frac{1}{2}} + B_0(i,j)\left(E_{zi,j}^{n} - E_{zi,j+1}^{n}\right)$$

and

$$H_{y\,i+\frac{1}{2},j}^{\,n+\frac{1}{2}} = \frac{1}{1+\frac{\Delta t}{2}\frac{\sigma^{\rm m}_{i+\frac{1}{2},j}}{\mu_{i+\frac{1}{2},j}}}\left[\left(1-\frac{\Delta t}{2}\frac{\sigma^{\rm m}_{i+\frac{1}{2},j}}{\mu_{i+\frac{1}{2},j}}\right)H_{y\,i+\frac{1}{2},j}^{\,n-\frac{1}{2}} + \frac{\Delta t}{\Delta x}\frac{1}{\mu_{i+\frac{1}{2},j}}\left(E_{zi+1,j}^{\,n} - E_{zi,j}^{\,n}\right)\right]$$
$$= A_1(i,j)H_{y\,i+\frac{1}{2},j}^{\,n-\frac{1}{2}} + B_1(i,j)\left(E_{zi+1,j}^{\,n} - E_{zi,j}^{\,n}\right) \qquad (3.45b)$$

give the advancement of the magnetic fields to time $t = (n + \frac{1}{2})\Delta t$ in terms of the magnetic and electric fields in the two earlier time steps, $t = (n - \frac{1}{2})\Delta t$ and $t = n\,\Delta\,t$, respectively. For the electric field [12]

$$E_{zi,j}^{\,n+1} = \frac{1}{1+\frac{\Delta t}{2}\frac{\sigma^{\rm e}_{i,j}}{\varepsilon_{i,j}}}\left[\left(1-\frac{\Delta t}{2}\frac{\sigma^{\rm e}_{i,j}}{\varepsilon_{i,j}}\right)E_{zi,j}^{\,n} + \frac{\Delta t}{\Delta x}\frac{1}{\varepsilon_{i,j}}\left(H_{y\,i+\frac{1}{2},j}^{\,n+\frac{1}{2}} - H_{y\,i-\frac{1}{2},j}^{\,n+\frac{1}{2}}\right)\right.$$
$$\left.+\frac{\Delta t}{\Delta y}\frac{1}{\varepsilon_{i,j}}\left(H_{x\,i,j-\frac{1}{2}}^{\,n+\frac{1}{2}} - H_{x\,i,j+\frac{1}{2}}^{\,n+\frac{1}{2}}\right)\right] = A_2(i,j)E_{zi,j}^{\,n} + B_2(i,j)\left(H_{y\,i+\frac{1}{2},j}^{\,n+\frac{1}{2}} - H_{y\,i-\frac{1}{2},j}^{\,n+\frac{1}{2}}\right) \qquad (3.45c)$$
$$+ C_2(i,j)\left(H_{x\,i,j-\frac{1}{2}}^{\,n+\frac{1}{2}} - H_{x\,i,j+\frac{1}{2}}^{\,n+\frac{1}{2}}\right)$$

gives the advancement of the electric fields to time $t = (n + 1)\,\Delta t$ in terms of the magnetic and electric fields in the two earlier time steps, $t = (n + \frac{1}{2})\Delta t$ and $t = n\,\Delta t$, respectively. Applying (3.45a) and (3.45b) advances the fields in time so that (3.45c) can then be applied to further advance them in time. This advancement is followed again by the time advancement provided by (3.45a) and (3.45b), etc. The cycle is repeated over and over. All that is needed to complete the description of the simulation process are boundary conditions for starting and ending the simulation. These conditions are discussed later.

Notice that the field coefficients—$A_l(i,j)$ and $B_l(i,j)$ for $l = 0, 1, 2$ and $C_2(i,j)$—in (3.45) do not depend on the update time. These coefficients need only be calculated and stored once in the simulation program. The electric and magnetic fields at a given time step are calculated from the fields of the two previous time steps, and these are the only fields that need to have been previously stored. These storage factors facilitate the implementation of the finite-difference time-domain method as compared to some of the other methods available in simulation approaches to photonics problems. It should also be noted that in computing the field coefficients for (3.45) care must be taken in choosing the values of Δt, Δx, and Δy for the simulation. The time step must be set so as to accurately represent the time variation of the mode of interest in the system with the maximum frequency. The time step should be much less than the period of the maximum frequency mode to be simulated but not so small as to excessively slow the development of the simulation in time. Similarly, the spatial mesh must provide a reasonably accurate representation of the spatial topography of the excitations of the system. It must not be so fine, however, as to excessively slow the simulation.

In the matter of boundary conditions, there are two general considerations [12]. (i) To start the simulation process going, a source of electromagnetic fields must be introduced into the system. It is the response of the system to the electromagnetic radiation from the source that reveals the properties of the system. Discussions here will be provided to determine the interaction of the system with an incident plane wave. This is a common consideration for the treatment of scattering problems. (ii) A second point involves the finite spatial extent of the simulation region. This is a fundamental limitation on computer simulations arising from the finite memory of a computer. Simulations involve the scattering of radiation from composite systems contained in a finite spatial region provided by the computer memory. Eventually the scattered radiation of the system will reach the outer boundary of the finite spatial region of the simulation. In order that the simulation gives a good approximation of the scattering of the composite system in infinite space, the radiation that reaches the outer boundary of the simulation must be absorbed rather than reflected back to again interact with the composite system. Effective ABCs at the outer edges of the finite simulation region are required for accuracy.

The bounded spatial region in which the computer simulation is run is usually divided into two regions (see figure 3.4). The bounded total region of the simulation is rectangular. Within this bounded region is an inner rectangular region containing

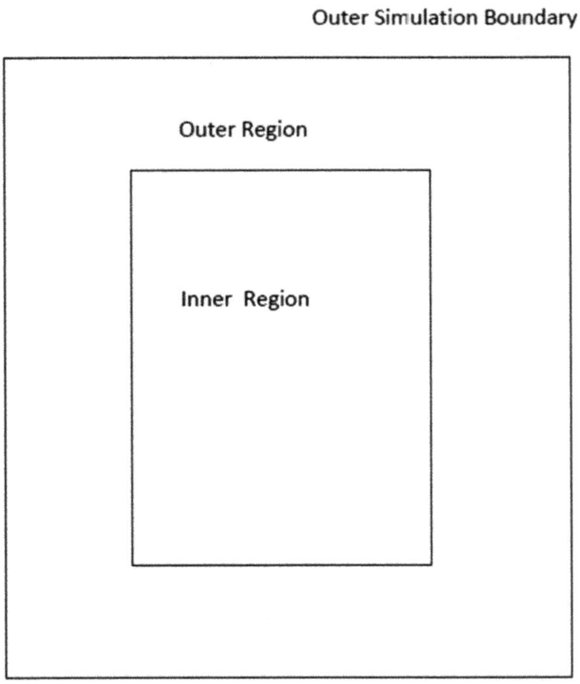

Figure 3.4. Simulation inner and outer regions and the outer boundary of the simulation.

composite materials that are to scatter the incident plane waves. This region contains both the incident and scattered waves. Outside the inner rectangular region is a region of uniform permittivity and permeability in which only the scattered waves from the inner rectangle are allowed to travel. Boundary conditions are set at the edges of the inner rectangular region in order to introduce the incident wave into inner rectangular region of the simulation. The incident waves are introduced by applying appropriate boundary conditions on all four sides of the inner rectangle. In addition, the boundary conditions at the inner region allow only the scattered wave to leave the inner region and travel to the outer edges of the total simulation region. Once the scattered wave leaves the inner region additional considerations are needed. For an infinite system the surfaces of the outer rectangular region would extend to infinity and the system would reduce to that which the simulation is meant to approximate. However, since the simulation is run in the bounded region of the outer rectangle, any scattered waves reaching the outer boundary of the simulation must be absorbed. These are treated by additional boundary conditions at the outer edge of the total simulation region.

To simulate a plane wave incident on the inner rectangular region, consider one of the sides of the inner rectangle which has an incident plane wave introduced as a boundary condition. Within the inner region the total field is composed of the incident wave and the scattered wave. The total field in the inner region evolves in time by (3.45). Outside the inner region only the scattered wave exists and it also evolves in time by (3.45). Boundary conditions are required at the side of the inner rectangle to match the two different types of solutions inside and outside the inner rectangle so that it appears that the incident plane wave is arising at the edge. The boundary conditions must be applied at each edge of the inner rectangular region.

As an example, consider the boundary treatment at the lower edge of the inner rectangle. Take the lower edge to be located at $y = j_l \Delta y$. The fields on and above this edge are total fields while those below the edge are only scattered fields. Applying (3.45c) to the electric fields on the edge gives the time evolution equation [12]

$$E_{z\ i,j_l}^{T\ n+1} = A_2(i,j_l)E_{z\ i,j_l}^{T\ n} + B_2(i,j_l)\left(H_{y\ i+\frac{1}{2},j_l}^{T\ n+\frac{1}{2}} - H_{y\ i-\frac{1}{2},j_l}^{T\ n+\frac{1}{2}}\right)$$

$$+ C_2(i,j_l)\left(H_{x\ i,j_l-\frac{1}{2}}^{S\ n+\frac{1}{2}} - H_{x\ i,j_l+\frac{1}{2}}^{T\ n+\frac{1}{2}}\right) + C_2(i,j_l)H_{x\ i,j_l-\frac{1}{2}}^{I\ n+\frac{1}{2}} \qquad (3.46a)$$

where superscripts have been introduced on the fields to indicate the total, scattered, and incident fields. The incident fields have been separated out as being needed in addition to the scattered and total fields. These incident fields are fed into (3.46a). In this form (3.46a) is a boundary condition term which introduces the incident wave into the system and only passes a scattered wave outside the inner region. The scattered and total fields of the equation are those evolving in time from the incident wave. In addition, the magnetic field equations at the lower edge of the inner rectangle also need to be rewritten. At the same lower edge of the rectangle (3.45a) becomes

$$H_{x\;i,j_l-\frac{1}{2}}^{S\;n+\frac{1}{2}} = A_0(i, j_l - 1)H_{x\;i,j_l-\frac{1}{2}}^{S\;n-\frac{1}{2}} + B_0(i, j_l - 1)\left(E_{z\;i,j_l-1}^{S\;n} - E_{z\;i,j_l}^{T\;n}\right)$$
$$+ B_0(i, j_l - 1)E_{z\;i,j_l}^{I\;n}. \tag{3.46b}$$

Again the incident fields are introduced or fed into the simulation by the last term on the right-hand side of the equation. At the lower edge, however, (3.45b) remains unchanged as it only couples terms along the x-axis, i.e.,

$$H_{y\;i+\frac{1}{2},j_l}^{T\;n+\frac{1}{2}} = A_1(i, j_l)H_{y\;i+\frac{1}{2},j_l}^{T\;n-\frac{1}{2}} + B_1(i, j_l)\left(E_{z\;i+1,j_l}^{T\;n} - E_{z\;i,j_l}^{T\;n}\right). \tag{3.46c}$$

Equations (3.46a) and (3.46b) are seen to couple the scattered fields below and the total fields at and above the lower edge to one another. At the same time they both introduce an incident wave source term into the simulation equations. This additional incident wave source term is needed to generate the evolution of the scattered and total fields in the regions below and above the edge of the rectangle. Considerations of this type are also needed along the other three sides of the rectangular inner region, with some special considerations at the vertices of the rectangular region. The resulting solution leaves the scattered wave moving in the region outside of the inner rectangle containing the composite medium. For further details of these boundary conditions the reader is referred to [12].

The final set of boundary conditions is those at the outer edges of the entire simulation region [12]. At the outer edge the scattered radiation must be absorbed so that it does not reflect back into the scattering system. There have been a number of methods proposed to achieve an approximation of this effect at the outer edge of the simulation. There are two basic ideas upon which most approaches are based.

The PML approach is based on placing at the outer edges of the simulation a layer which has dissipative electrical and magnetic properties [12]. These properties are modeled by the electrical conductivity, σ^{e}, and the magnetic conductivity, σ^{m}. To understand the basic ideas of the method, the ideas will be developed for the motion of TM waves. Consider the TM equations given in (3.38) for a planar interface between a medium without electrical or magnetic conductivity and a medium characterized by the electrical conductivity, σ^{e}, and the magnetic conductivity, σ^{m}. Aside from their conductivity properties the two media are identical. To begin with, we determine the condition on σ^{e} and σ^{m} under which a TM wave at normal incidence to the planar interface has no reflected wave from the interface. The interface is then effectively absent as it gives no reflected wave.

For the calculation take the planar interface to be the y–z plane so that the nonconductive medium is in the region $x < 0$ and the conductive medium is in the region $x \geqslant 0$. From (3.38) a TM wave propagating along the x-axis in the nonconductive medium has the form

$$E_z^{\mathrm{nc}} = E_{z0}^{\mathrm{nc}}\mathrm{e}^{\mathrm{i}(kx-\omega t)} \tag{3.47a}$$

$$H_y^{\mathrm{nc}} = H_{y0}^{\mathrm{nc}}\mathrm{e}^{\mathrm{i}(kx-\omega t)}. \tag{3.47b}$$

Substituting (3.47) into (3.38) for the nonconductive case gives

$$\frac{\omega}{k} = \frac{1}{\sqrt{\varepsilon\mu}} \tag{3.48a}$$

and

$$H_{y0}^{\mathrm{nc}} = -\frac{\varepsilon}{\sqrt{\varepsilon\mu}} E_{z0}^{\mathrm{nc}} \tag{3.48b}$$

respectively, for the dispersion relation and the relationship between the electric and magnetic field amplitudes. In the conductive medium the form of a TM wave propagating along the x-axis is given by

$$E_z^{\mathrm{c}} = E_{z0}^{\mathrm{c}} \mathrm{e}^{\mathrm{i}(kx-\omega t)} \tag{3.49a}$$

and

$$H_y^{\mathrm{c}} = H_{y0}^{\mathrm{c}} \mathrm{e}^{\mathrm{i}(kx-\omega t)}. \tag{3.49b}$$

Substituting (3.49) into (3.38) for the conductive case we find

$$k^2 = \varepsilon\mu\left(\omega + \mathrm{i}\frac{\sigma^{\mathrm{e}}}{\varepsilon}\right)\left(\omega + \mathrm{i}\frac{\sigma^{\mathrm{m}}}{\mu}\right) \tag{3.50a}$$

$$H_{y0}^{\mathrm{c}} = -\varepsilon \frac{\omega + \mathrm{i}\frac{\sigma^{\mathrm{e}}}{\varepsilon}}{k} E_{z0}^{\mathrm{c}} \tag{3.50b}$$

for the dispersion relation and the relation between the electric and magnetic field amplitudes.

In the case that

$$\frac{\sigma^{\mathrm{e}}}{\varepsilon} = \frac{\sigma^{\mathrm{m}}}{\mu} \tag{3.51}$$

the solutions at the $x = 0$ interface between the nonconductive medium in (3.47) and (3.48) and the conductive medium in (3.49) and (3.50) reduce to one another, i.e., $E_{z0}^{\mathrm{nc}} = E_{z0}^{\mathrm{c}}$ and $H_{y0}^{\mathrm{nc}} = H_{y0}^{\mathrm{c}}$. Consequently, the conductive and nonconductive media are perfectly matched at their interface, and a reflected wave is not generated at the interface. Inside the conductive medium, however, from (3.50a) the wave vector is seen to be complex. This causes the wave to decay during its propagation along the x-axis in the absorbing medium and to be effectively removed from the system.

The PML method is applied by assuming a conductive layer at the outer boundary of the simulation. The conductivities of the outer layer are chosen such that (3.51) holds between the medium carrying the scattered wave to the outer boundary of the simulation and the outer conductive (PML) layer. While the conditions of (3.51) are obtained for a perfect matching at normal incidence they are approximated conditions for matching at off normal incidence, and some further considerations are addressed in the PML to extend the range of incident angles that have well matched boundary conditions [12]. For the TM case this involves writing the electric field as the sum of two subfields and rewriting the equation for the time derivative of the electric field as two separate equations for the time derivatives of the postulated subfields. The reader is referred to the literature for more details and discussions of the accuracy and of the general use of these boundary conditions. These references also show effective methods for discretizing the PML equations and setting the dimensions of the PML region and indicate that the PML is in general a great improvement over other ABC methods [12].

Another class of approaches to creating an absorbing outer boundary is to place on the outer boundary of the simulation region a layer of medium which only allows waves to propagate away from the inner scattering region. In this layer, waves can only move out of the simulation as there are no solutions for waves moving into the simulation region. The boundary layer is also chosen to have dielectric and magnetic properties which match those of the medium in which the scattered wave propagates to the simulation boundary. Consequently, no wave is reflected from the interface of the medium in which the scattered wave propagates and the layer at the outer boundary. The layer at the outer boundary differs from the PML as it does not employ a dissipative electric and magnetic conductivity. Rather it is a directional medium.

To understand the operation of the outer boundary layer, consider the TM system in (3.38). These equations can be rewritten as the second order partial differential equations [12]

$$\left(\frac{\partial^2}{\partial x^2} + \frac{\partial^2}{\partial y^2} - \varepsilon\mu\frac{\partial^2}{\partial t^2}\right)\left\{\begin{matrix} H_x \\ H_y \\ E_z \end{matrix}\right\} = 0. \tag{3.52}$$

The resulting equation describes the propagation of the plane wave solutions in the region outside the inner region as the scattered waves propagate towards the simulation outer boundary. The reflection of the scattered waves at the outer boundary of the simulation needs to be prevented.

To see how this might be accomplished we will look at one of the edges of the outer boundary of the simulation. Consider creating an absorbing layer at the right-hand boundary of the simulation with the same dielectric and magnetic properties as the medium in which the scattered waves approach the simulation boundary. From (3.52) an operator for the x-component of the wave vector of the waves moving in the scattering region is

$$-\mathrm{i}\frac{\partial}{\partial x} = \pm\left[\frac{\partial^2}{\partial y^2} - \varepsilon\mu\frac{\partial^2}{\partial t^2}\right]^{\frac{1}{2}}. \tag{3.53}$$

The operator in (3.53) has two types of eigenvalue solutions for the two signs of the square root on the right-hand side of the equations. The positive sign gives the solutions moving towards the right-hand boundary of the simulation and the negative sign gives the solutions moving away from the right-hand boundary of the simulation. Combining the solutions for both signs reproduces the entire plane wave solution set of (3.52).

In the proposed boundary layer at the outer right-hand edge of the simulation, the layer medium is chosen so that the wave propagation in the layer medium is described by

$$\mathrm{i}\frac{\partial}{\partial x} = -\left[\frac{\partial^2}{\partial y^2} - \varepsilon\mu\frac{\partial^2}{\partial t^2}\right]^{\frac{1}{2}} \tag{3.54}$$

rather than by (3.52) or (3.53). This equation only allows motion of waves in the boundary layer which is out of the system. The medium in (3.54) has the same dielectric and magnetic properties as the medium described by (3.52) and (3.53) so that there should be no reflection at the interface between the layer and its adjacent medium. However, only waves of positive x-component of the wave vector reach the layer as they travel from the inner scattering region, and only these are passed through to the boundary layer. No waves of negative x-component can ever reach the right outer boundary layer nor can they originate in the layer which does not support them. Similar considerations must be made at the remaining edges and corners of the outer boundary of the simulation region.

The problem with implementing (3.54) in the outer boundary layer of the simulation is how to treat the square root in a simplified manner for the development of the numerical simulation. This has given rise to a number of approximation methods which replace the square root by a power law expansion. These will now be noted. To understand the process, consider applying the method to study radiation at the particular frequency, ω. Equation (3.54) then becomes

$$\mathrm{i}\frac{\partial}{\partial x} = -\left[\varepsilon\mu\omega^2 + \frac{\partial^2}{\partial y^2}\right]^{\frac{1}{2}}. \tag{3.55}$$

A common type of approximation for the square root in (3.55) is to make a Taylor expansion. Retaining the first two terms in such an expansion gives

$$\mathrm{i}\frac{\partial}{\partial x} = -\sqrt{\varepsilon\mu}\,\omega\left[1 + \frac{1}{2}\frac{1}{\varepsilon\mu\omega^2}\frac{\partial^2}{\partial y^2}\right]. \tag{3.56}$$

This, in turn, can be written as

$$-\mathrm{i}\omega\frac{\partial}{\partial x} = \sqrt{\varepsilon\mu}\,\omega^2 + \frac{1}{2}\frac{1}{\sqrt{\varepsilon\mu}}\frac{\partial^2}{\partial y^2}. \tag{3.57}$$

Returning to operator form by representing the frequency as a time derivative (3.57) becomes

$$\frac{\partial^2}{\partial x \partial t} = -\sqrt{\varepsilon\mu}\frac{\partial^2}{\partial t^2} + \frac{1}{2}\frac{1}{\sqrt{\varepsilon\mu}}\frac{\partial^2}{\partial y^2} \tag{3.58}$$

so that the waves propagating in this directional medium then satisfy

$$\left(\frac{\partial^2}{\partial x \partial t} - \frac{1}{2}\frac{1}{\sqrt{\varepsilon\mu}}\frac{\partial^2}{\partial y^2} + \sqrt{\varepsilon\mu}\frac{\partial^2}{\partial t^2}\right)\begin{Bmatrix} H_x \\ H_y \\ E_z \end{Bmatrix} = 0. \tag{3.59}$$

Equation (3.59) only describes waves with positive x-components of the wave vectors and is a linear equation which can be used as a basis for of the finite-difference time-domain method in the layer at the outer boundary. This involves the discretization of the space and time derivatives and the evolution of (3.59) into a set of difference equations. Equation (3.59) is an approximation which is only as good as the approximation made to the square root factor. Here a Taylor expansion has been used, but a number of different methods including Pade approximation methods and expansions in orthogonal polynomials have been suggested to improve the approximation.

The reader is again referred to the literature for details of the accuracy and implementation of the method based on a directional layer [12].

3.3.2 Method of moments

The method of moments is used to treat the frequency-dependent properties of a system. In this case there is no time variation to deal with, but the spatial behavior of the frequency modes of the system are the focus of the consideration [14]. The treatment of the wave functions is preformed numerically by a discretization in space of the wave function equations at each frequency. As with the finite-difference time-domain method the spatial extent of the system treated is limited by the storage capacity of the computer running the simulation. In the case of the method of moments the limit does not involve a time variable but only involves the spatial size of the sample of the system treated. The sample size is restricted to be of a finite extent and requires that the simulated system be large enough to approximate the behavior of the infinite system it models. Excitation wave functions are obtained for each frequency by starting from an integral equation formulation of the problem which is generated from the Maxwell equations, discretizing the integral equations into the form of matrix equations, and solving these using techniques of linear algebra.

To understand the basics of the implementation of the method of moments consider a two-dimensional scattering problem for the case in which TM waves traveling in free space scatter from perfect conductors [14]. The waves propagate in the x–y plane with electric fields polarized along the z-axis, and the perfect conductors are translational invariant along the z-axis. The Helmholtz equations for the electromagnetic fields of frequency, ω, outside the perfect conductors are given by

$$\left(\frac{\partial^2}{\partial x^2} + \frac{\partial^2}{\partial y^2} + \varepsilon_0\mu_0\omega^2\right)\begin{Bmatrix} H_x \\ H_y \\ E_z \end{Bmatrix} = 0. \tag{3.60}$$

These fields are of the form

$$E_z(x, y, \omega) = E_z^{\mathrm{I}}(x, y, \omega) + E_z^{\mathrm{S}}(x, y, \omega) \tag{3.61}$$

where the superscripts I and S, respectively, refer to the incident and scattered fields. Insider the perfect conductors the total electric field is zero so that at the surface of the perfect conductors

$$E_z^{\mathrm{I}}(x, y, \omega)_{\mathrm{Surface}} = -E_z^{\mathrm{S}}(x, y, \omega)_{\mathrm{Surface}}. \tag{3.62}$$

The next step in the formulation is to replace the above discussed scattering problem of the perfect conducting system by an equivalent antenna problem. The antenna problem is chosen so that the antenna radiates fields that are the same as those generated as scattered waves in the scattering problem. The idea of the equivalent problem is to determine a set of surface currents on the scattering surfaces that would radiate the same scattered wave as is developed in the perfect conductor scattering problem. In the equivalent antenna problem the scattering surfaces are no longer perfect conductions but rather they are surfaces containing surfaces currents. For the moment assume that the surface currents are known and are of the form, $J_{\mathrm{Surface}}(x, y, \omega)$. Later discussions will be provided on how to arrive at the appropriate surface currents for the equivalent antenna problem.

The solutions for the waves radiated by $J_{\mathrm{Surface}}(x, y)$ can be written in terms of the Green function of the operator in (3.60). The Green function is obtained as a solution of

$$\left[\frac{\partial^2}{\partial x^2} + \frac{\partial^2}{\partial y^2} + \varepsilon_0\mu_0\omega^2\right]G(x, y; x', y', \omega) = -\delta(x - x')\delta(y - y') \tag{3.63}$$

and for an unbound infinite space with radiation boundary conditions the Green function is written in terms of Hankel functions. The formal solutions for the scattered wave in terms of the retarded Green function and surface current is then

$$E_z^{\mathrm{S}}(x, y, \omega) = \mathrm{i}\mu_0\omega \int_{\mathrm{Surfaces}} \mathrm{d}x'\mathrm{d}y' G(x, y; x', y', \omega) J_{\mathrm{Surface}}(x', y', \omega). \tag{3.64}$$

Evaluating this for the fields at the scattering surfaces

$$\begin{aligned} E_Z^S(x, y, \omega)_{\mathrm{Surface}} &= -E_z^I(x, y, \omega)_{\mathrm{Surface}} \\ &= \mathrm{i}\mu_0\omega \int_{\mathrm{Surfaces}} \mathrm{d}x'\mathrm{d}y' G(x, y; x', y', \omega) J_{\mathrm{Surface}}(x', y', \omega), \end{aligned} \tag{3.65}$$

where the first equality is due to the zero of the total electric field, composed of incident and scattered waves, at the perfect conducting surface. The incident wave is of a known form (i.e., it is often taken to be a plane wave) so that from the second equation in (3.65)

$$E_z^{\mathrm{I}}(x, y, \omega)_{\mathrm{Surface}} = -\mathrm{i}\mu_0\omega \int_{\mathrm{Surfaces}} \mathrm{d}x'\mathrm{d}y' G(x, y; x', y', \omega) J_{\mathrm{Surface}}(x', y', \omega) \tag{3.66}$$

relates the known incident field to the unknown surface current which, from (3.65), would generate the scatter field at the surface. To obtain the antenna current of the equivalent antenna problem then requires (3.66) to be inverted for the surface current. Once this is done the resulting surface current is then used in (3.64) to determine the scattered wave everywhere in space.

Inverting (3.66) for the surface current is done by rewriting (3.66) as a matrix equation which is then solved for the surface current. A common method to do this is to cover the conducting surfaces with a finite array of points. At each array point on the surface, $\vec{r}_i = x_i\hat{i} + y_i\hat{j}$, a function, $f_i(x, y)$, is defined which is one at the array point, $\vec{r}_i = x_i\hat{i} + y_i\hat{j}$, and nonzero only over a finite region of space approaching its nearest neighbor points on the surface. In addition,

$$f_i(\vec{r}_j) = \delta_{i,j}. \tag{3.67}$$

This is done so that the entire surface is covered by the above defined functions. The surface current is then written in terms of these functions as

$$J_{\text{Surface}}(x, y, \omega) = \sum_i a_i f_i(\vec{r}). \tag{3.68}$$

Here it is seen that the number of array points, $\vec{r}_i = x_i\hat{i} + y_i\hat{j}$, must be enough to accurately represent the surface current and the functions, $f_i(x, y)$, also must be such as to adequately give a representation of the surface currents.

Substituting (3.68) for the surface current in (3.66) gives

$$\begin{aligned} &E^{\text{I}}(x, y, \omega)_{\text{Surface}} \\ &\quad = -\mathrm{i}\mu_0\omega \int_{\text{Surfaces}} \mathrm{d}x'\mathrm{d}y' G(x, y; x', y', \omega)\sum_j a_j f_j(x', y'). \end{aligned} \tag{3.69}$$

Multiplying both sides of (3.69) by $f_i(x, y)$ and integrating over x and y generates a matrix equation

$$b_i = \sum_j M_{i,j} a_j \tag{3.70}$$

where

$$M_{i,j} = -\mathrm{i}\mu_0\omega \int_{\text{Surface}} \mathrm{d}x\mathrm{d}y\mathrm{d}x'\mathrm{d}y' f_i(x, y)G(x, y; x', y', \omega)f_j(x', y') \tag{3.71}$$

and

$$b_i = \int_{\text{Surface}} \mathrm{d}x\mathrm{d}y f_i(x, y)E^{\text{I}}(x, y, \omega)_{\text{Surface}}. \tag{3.72}$$

Equation (3.70) is to be solved for the unknown in a_j terms of the known b_i. Once the a_j are determined the surface current is obtained from (3.68) and the scattered field everywhere from (3.64).

The method of solution for the scattered fields, then, relies on the inversion of the matrix equation in (3.70). For most systems of interest the matrix equations involve large matrices. This can be a complication, but a simplification occurs in many problems in which the matrix can be written in the form of a positive symmetric matrix. In this case a solution is greatly facilitated using the conjugate gradient method. Consequently, the conjugate gradient method is now briefly outlined [14].

The matrix equation in (3.70) written in operator form

$$b = Ma \tag{3.73}$$

can be symmetrized by using the matrix transpose to rewrite (3.73) as

$$M^{\mathrm{T}}b = M^{\mathrm{T}}Ma \tag{3.74}$$

where $M^{\mathrm{T}}M$ is a positive symmetric matrix. This can then be rewritten as

$$c = Aa \tag{3.75}$$

where $A = M^{\mathrm{T}}M$ is a known matrix and $c = M^{\mathrm{T}}b$ is a known vector.

The conjugate gradient method generates the solution to (3.75) by using the fact that the solution of (3.75) minimizes the nonlinear function

$$g(a) = \frac{1}{2}a^{\mathrm{T}}Aa - a^{\mathrm{T}}c. \tag{3.76}$$

The idea of the method is to generate a set of orthogonal vectors which span the vector space in which the solution of (3.75) is defined. As a result, the solution of (3.75) is then a linear combination of these orthogonal vectors. The orthogonal set is generated during the process of iteratively searching for the solution of (3.75) where, during the iterative process, the method systematically generates closer and closer successive approximations to the solution of the equation. This is a much quicker process than direct matrix inversion.

To see how this works, consider that the set of orthogonal vectors in question is $\{s_i\}$ with orthogonality properties defined by the inner product $s_i^{\mathrm{T}}As_j$ where

$$s_i^{\mathrm{T}}As_j = 0 \quad \text{for} \quad i \neq j \tag{3.77}$$

and $s_i^{\mathrm{T}}As_i$ is nonzero for $i = j$. The solution of (3.75) then has the form

$$a = \sum_i \alpha_i s_i = \sum_i a_i \tag{3.78}$$

which when substituted into (3.75) gives

$$c = \sum \alpha_i A s_i. \tag{3.79}$$

From (3.79) and (3.77), the coefficients in (3.78) are given by

$$\alpha_i = \frac{s_i^{\mathrm{T}}c}{s_i^{\mathrm{T}}As_i}. \tag{3.80}$$

A set of orthogonal vectors $\{s_i\}$ must be generated for the solution given in (3.78) and (3.80). This is achieved by considering the negative of the gradient of (3.76)

$$-g'(x) = c - Ax \tag{3.81}$$

evaluated at the vectors x in the vector space defined by $\{s_i\}$. At the solution of (3.75) $g'(a) = 0$, but at other points (3.81) is a nonzero vector defining a direction in the vector space. The idea is to make an approximation of the solution by minimizing $g(x)$ along the direction defined from (3.81) by an initial guess for the solution. Consequently (3.81) is used to generate the vectors $\{s_i\}$ and the procedure is somewhat similar to Gram–Schmidt orthogonalization [5].

To see how the set $\{s_i\}$ is generated, start by guessing that $a \approx \tilde{a}_0 = 0$ is the solution of (3.75) and evaluate (3.81) at $x = \tilde{a}_0 = 0$. Substituting this in (3.81) gives

$$s_0 = c - A\tilde{a}_0 = c \tag{3.82}$$

which is taken as the first vector spanning the space of solutions. Then from (3.80)

$$\alpha_0 = \frac{s_0^{\mathrm{T}} c}{s_0^{\mathrm{T}} A s_0} \tag{3.83}$$

and from (3.78) in this approximation an improved solution along the vector s_0 is

$$a \approx a_0 = \alpha_0 s_0. \tag{3.84}$$

For the next step, the vector in (3.84) is taken as the initial guess for a solution of (3.75). Taking $a = a_0$ in (3.75) generates the remainder (i.e., the measure of how bad the initial guess a_0 is) given by

$$r_1 = c - Aa_0. \tag{3.85a}$$

The remainder in (3.85a) is now used to make the second orthogonal vector,

$$s_1 = r_1 - \frac{s_0^{\mathrm{T}} A r_1}{s_0^{\mathrm{T}} A s_0} s_0 \tag{3.85b}$$

of the orthogonal set. Notice that the second term in (3.85b) is included to assure that s_0 and s_1 are orthogonal. Using this vector in (3.78) and (3.80) gives $\alpha_1 = \frac{s_1^{\mathrm{T}} c}{s_1^{\mathrm{T}} A s_1}$, and after this iteration an approximate solution is $a \approx a_0 + a_1 = \alpha_0 s_0 + \alpha_1 s_1$.

At the ith step of the iteration process, the approximate solution from the previous iterations is $a \approx a_0 + a_1 + \cdots + a_{i-1}$ and a substitution of this approximation in (3.75) generates the new remainder vector

$$r_i = c - A(a_0 + a_1 + \cdots + a_{i-1}) \tag{3.86a}$$

from which the new orthogonal vector is

$$s_i = r_i - \sum_{k<i} \frac{s_k^{\mathrm{T}} A r_i}{s_k^{\mathrm{T}} A s_k} s_k. \tag{3.86b}$$

Note that the second term on the right-hand side of (3.86b) assures that s_i is orthogonal to the earlier generated set of $\{s_k\}$. The new improvement on the solution is

$$a_i = \alpha_i s_i \tag{3.87}$$

with α_i obtained from (3.80) so that $a \approx a_0 + a_1 + \cdots + a_i$ is the new approximate solution. During this procedure an orthogonal set $\{s_i\}$ is developed, expressed in terms of the remainder $r_i = c - A(a_0 + a_1 + \cdots + a_{i-1})$ generated at each iterative step. The difference between the guess and the actual solution is seen to gradually be reduced. The process can be stopped when the remainder is sufficiently decreased. This means that the entire matrix does not need to be inverted, the matrix inversion being a much lengthier process to develop.

3.3.3 Finite element method

The final simulation technique to be discussed is the finite element method [15]. This methodology is very general and usually associated with time-independent problems or frequency-dependent problems.

To understand the basic implementation of the method, consider the treatment of the two-dimensional Helmholtz equation for the determination of its modes. The equation is of the basic form

$$\left(\frac{\partial^2}{\partial x^2} + \frac{\partial^2}{\partial y^2}\right)\varphi + k^2\varphi = 0 \tag{3.88}$$

where for simplicity the case in which $\varphi(x, y)$ is a scalar is treated. The problem is defined over a finite region of space and has boundary conditions on the edges of the spatial region. Equation (3.88) is often encountered in the propagation of waves in dielectric or metallic waveguides. In addition, when $k = 0$, it reduces to the two-dimensional Laplace equation studied in electrostatics and potential flow.

Equation (3.88) is obtained from minimizing the functional [15]

$$\Gamma(\varphi) = \frac{1}{2}\iint \mathrm{d}x\mathrm{d}y\left[\left(\frac{\partial\varphi}{\partial x}\right)^2 + \left(\frac{\partial\varphi}{\partial y}\right)^2 - k^2\varphi^2\right] \tag{3.89}$$

and this forms the basis of the finite element approach. The idea of the finite element method is to minimize the functional in (3.89) numerically by converting (3.89) into an algebraic form which is then minimized. This is done by breaking the two-dimensional x–y space up into little triangles that do not overlap one another but nonetheless cover the region over which the two-dimensional problem is defined. The vertices of each triangle are labeled by $i = 1, 2, 3$ and the triangles are themselves numbered $t = 1, 2, \ldots, N$ so that N is the total number of triangles in the covering set.

In this notation the potential in the region defined by the tth triangle is approximated by the linear form

$$\varphi_t(x, y) = a_t + b_t x + c_t y. \tag{3.90}$$

Evaluating (3.90) at the three vertices, (x_i, y_i), of the tth triangle generates three linear equations for the coefficients a_t, b_t, c_t. The solution of these three equations gives

$$a_t = \sum_{i=1}^{3} a_{ti}\varphi_t(x_i, y_i) \tag{3.91a}$$

$$b_t = \sum_{i=1}^{3} b_{ti}\varphi_t(x_i, y_i) \tag{3.91b}$$

$$c_t = \sum_{i=1}^{3} c_{ti}\varphi_t(x_i, y_i) \tag{3.91c}$$

written in terms of the potentials at the vertices of the tth triangle. Substituting (3.91) into (3.90) we find

$$\varphi_t(x, y) = \sum_{i=1}^{3} (a_{ti} + b_{ti}x + c_{ti}y)\, \varphi_t\, (x_i, y_i). \tag{3.92}$$

This gives an approximation for the scalar function within the tth triangle in terms of the potentials at the corners of the triangle.

The form for $\varphi_t(x, y)$ in (3.92) for each triangle can now be substituted in (3.89). Upon integrating (3.89) over x and y then generates an algebraic form involving the $\varphi_t(x_i, y_i)$ which are to be determined from the minimization process. The space integrals in this process are easily done as they involve polynomials of the x and y coordinates over the entire two-dimensional region of space. The result of the space integral is an algebraic form in the $\varphi_t(x_i, y_i)$ which is then minimized. The resulting $\varphi_t(x_i, y_i)$ are used in (3.92) at each triangle to generated the scalar function $\varphi(x, y)$ throughout space.

3.4 Surfaces

Another class of systems exhibiting important properties arising from periodicity occurs in the study of surfaces. These include surfaces with applications as diffraction gratings for the processing of light incident on them and plasmonic surfaces which bind electromagnetic surface waves known as surface plasmons. The technological applications of both of these structures arise from the physics associated with the periodic variation in the mean plane of the surface. Though for gratings and periodic plasmonic surfaces the nature of the light in each and its interaction with the periodicity is quite different.

The first application of periodicity for the design of diffraction gratings has a long history in the field of optics [17]. Gratings have been used in the design of many types of spectrometers, monochrometers, lasers, and has displayed applications in pulse compression, multiplexing, data storage, telecommunication devices, etc [17]. These are mostly associated with the frequency dependence of the interaction of the gratings with light incident upon them.

On the other hand, the application of periodic surfaces in plasmonics is primarily focused on electromagnetic surface wave excitations [18–22]. These electromagnetic modes are bound to the surface, propagating in it mean plane. The existence of electromagnetic surface waves on an interface depends on the properties of the two materials joined at the interface and the geometry of the interface. Not all surfaces can support electromagnetic surface wave modes, but when they do exist on a surface, they can have a number of important applications. Surface waves have found applications in the technologies of surface enhanced Raman spectroscopy (SERS), near field microscopy, communications, sensing, etc [18–22].

In the following, first the basic theory of grating structures is discussed. This is followed by a treatment of surface waves. It is left to a later chapter to treat some of the properties of these systems arising upon the introduction of nonlinear optical media.

3.4.1 Basics of grating theory

To begin the treatment of the theory of gratings, a review is presented of the scattering of light from a planar interface between two dielectric media. The Maxwell equations for the discussions are taken to be of the form [9]

$$\nabla \cdot \vec{E} = 0,\ \nabla \times \vec{E} + \frac{1}{c}\frac{\partial \vec{B}}{\partial t} = 0 \tag{3.93}$$

$$\nabla \cdot \vec{B} = 0,\ \nabla \times \vec{B} - \frac{\mu\varepsilon}{c}\frac{\partial \vec{E}}{\partial t} = 0$$

where $\varepsilon > 0$ and $\mu > 0$ are, respectively, the linear permittivity and permeability. From the equations in this form the general solutions for propagating electromagnetic plane waves are then given by

$$\vec{E} = \vec{E}_0 e^{i\left(\vec{k}\cdot\vec{r} - \omega t\right)} \tag{3.94a}$$

$$\vec{B} = \frac{\sqrt{\mu\varepsilon}}{k}\vec{k} \times \vec{E}_0 e^{i\left(\vec{k}\cdot\vec{r} - \omega t\right)} \tag{3.94b}$$

where $\frac{\omega}{k} = \frac{c}{\sqrt{\mu\varepsilon}}$.

For the problem of the refraction of light at a plane interface between two media the geometry in figure 3.5 is considered. In terms of this schematic, light is incident from below the interface which is located at the plane defined by $z = 0$. Upon encountering the interface, the incident light is partially refracted into the region above the interface and partially reflected back into the region below the interface. Both the refracted and reflected waves propagate away from the surface.

In the region $z < 0$ the medium is described by permittivity and permeability $\varepsilon,\mu > 0$ and the plane wave solutions are of the form of an incident wave given by

$$\vec{E} = \vec{E}_0 e^{i\left(\vec{k}\cdot\vec{r} - \omega t\right)},\ \vec{B} = \frac{\sqrt{\mu\varepsilon}}{k}\vec{k} \times \vec{E}_0 e^{i\left(\vec{k}\cdot\vec{r} - \omega t\right)} \tag{3.95a}$$

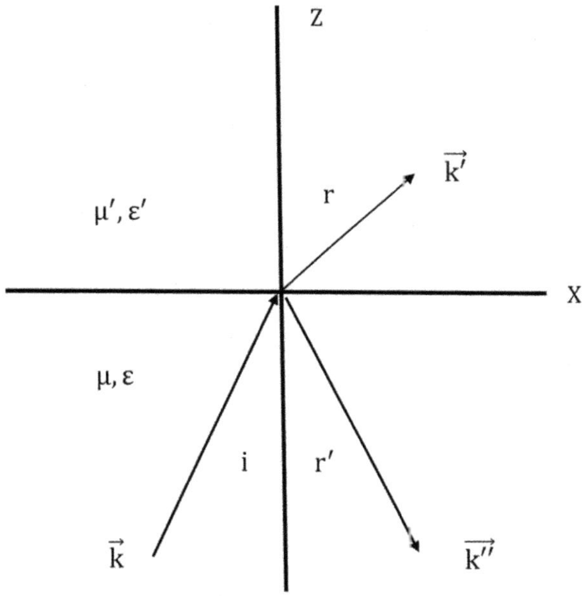

Figure 3.5. Refraction and reflection of an incident plane wave of light at the planar interface (located at $z = 0$) between a lower medium μ,ε and an upper medium μ',ε'. The angle of incidence i is measured clockwise from the negative z-axis, the angle of reflection r' is measured anti-clockwise from the negative z-axis, and the angle of refraction r is measured clockwise from the positive z-axis. The incident plane is the x–z plane.

where $\frac{\omega}{k} = \frac{c}{\sqrt{\mu\varepsilon}}$, and a reflected wave given by

$$\vec{E}^{''} = \vec{E}_0^{''} e^{i\left(\vec{k}^{''}\cdot\vec{r}-\omega t\right)},\ \vec{B}^{''} = \frac{\sqrt{\mu\varepsilon}}{k^{''}}\vec{k}^{''} \times \vec{E}_0^{''} e^{i\left(\vec{k}^{''}\cdot\vec{r}-\omega t\right)} \tag{3.95b}$$

where $\frac{\omega}{k^{''}} = \frac{c}{\sqrt{\mu\varepsilon}}$. In the region $z > 0$ the medium is described by the permittivity and permeability ε', $\mu' > 0$, and the electromagnetic refracted wave solutions are of the form [9]

$$\vec{E}' = \vec{E}_0' e^{i\left(\vec{k}'\cdot\vec{r}-\omega t\right)},\ \vec{B}' = \frac{\sqrt{\mu'\varepsilon'}}{k'}\vec{k}' \times \vec{E}_0' e^{i\left(\vec{k}'\cdot\vec{r}-\omega t\right)} \tag{3.95c}$$

where $\frac{\omega}{k'} = \frac{c}{\sqrt{\mu'\varepsilon'}}$. The solutions above and below the $z = 0$ surface are jointed together at the interface between the two regions by the electromagnetic boundary conditions. These require that the normal components of the vectors $\vec{D}$ and $\vec{B}$ are continuous and the tangential components of the vectors $\vec{E}$ and $\vec{H}$ are continuous.

Matching the solutions in (3.95) at the $z = 0$ interface for the case in which the incident wave E is polarized perpendicular to the page in figure 3.5 gives the two Fresnel relations [9]

$$\frac{E_0'}{E_0} = \frac{2n\cos i}{n\cos i + \frac{\mu}{\mu'}\sqrt{(n')^2 - n^2\sin^2 i}} \tag{3.96a}$$

$$\frac{E_0^{''}}{E_0} = \frac{n\cos i - \frac{\mu}{\mu'}\sqrt{(n')^2 - n^2 \sin^2 i}}{n\cos i + \frac{\mu}{\mu'}\sqrt{(n')^2 - n^2 \sin^2 i}} \tag{3.96b}$$

where $n = \sqrt{\mu\varepsilon}$ is the index of refraction of the lower medium and $n' = \sqrt{\mu'\varepsilon'}$ is the index of refraction of the upper medium. Alternatively for the case in which the incident wave E is polarized parallel to the page in figure 3.5 gives the two Fresnel relations [9]

$$\frac{E_0'}{E_0} = \frac{2nn'\cos i}{\frac{\mu}{\mu'}(n')^2\cos i + n\sqrt{(n')^2 - n^2 \sin^2 i}} \tag{3.96c}$$

$$\frac{E_0^{''}}{E_0} = \frac{\frac{\mu}{\mu'}(n')^2\cos i - n\sqrt{(n')^2 - n^2 \sin^2 i}}{\frac{\mu}{\mu'}(n')^2\cos i + n\sqrt{(n')^2 - n^2 \sin^2 i}}. \tag{3.96d}$$

In addition to the Fresnel relations, for $\vec{E}$ polarized either perpendicular or parallel to the page, the Snell's law relations [9]

$$n \sin i = n' \sin r = n \sin r' \tag{3.97}$$

holds for both polarizations. Snell's law follows directly from the requirements of complete translational symmetry in the plane of the surface so that $\left(\vec{k} \cdot \vec{r}\right)_{z=0} = \left(\vec{k}' \cdot \vec{r}\right)_{z=0} = \left(\vec{k}'' \cdot \vec{r}\right)_{z=0}$.

The relationships in (3.96) and (3.97) give a general description of the passage of electromagnetic plane waves between the two media. In this regard, it is interesting to note that for the case in which $r = \frac{\pi}{2}$ an incident wave does not pass from the lower media into the upper media. In fact, from (3.97) it follows that

$$k \sin i_0 = k' \tag{3.98a}$$

so that the wave vector in the upper medium points along the interface. This inability to pass between the two media can only occur provided that $n > n'$, and the angle of incidence for this to take place is given from (3.97) by

$$i_0 = \arcsin\frac{n'}{n}. \tag{3.98b}$$

Consequently, at incident angles for which $i > i_0$ only incident and reflected waves exist in the system.

The phenomenon is known as total internal reflection and is a fundamental limitation on the passage of light from one medium to another. To understand the nature of the plane wave solutions under the conditions of total internal reflection, consider that wave functions in the upper medium have an electromagnetic dispersion relation given by

$$\frac{\mu'\varepsilon'}{c^2}\omega^2 = k_x^2 + k_y^2 + k_z^2. \tag{3.99}$$

From (3.99) it is seen that in the case of wave vectors with $k_x^2 + k_y^2 \geqslant \frac{\mu'\varepsilon'}{c^2}\omega^2$ a real solution for ω is still possible provided that $k_z^2 \leqslant 0$. These are the conditions of total internal reflection where at i_0 the relationship $k'^2 = k_x^2 + k_y^2 = \frac{\mu'\varepsilon'}{c^2}\omega^2$ and for $i > i_0$ the inequality takes over. In this region the wave function solutions in (3.95c) are seen to exponentially decay with distance from the interface in the upper medium. The form of the wave function below the interface, however, remains in the form of (3.95a) and (3.59b) as a composite of incident and reflected bulk propagating waves traveling to and from the interface. In this regard, the total internal reflected modes are not bound to the interface between the two media but remain bulk waves in at least the lower medium.

3.4.1.1 Periodic surface profile

Consider now the case in which a one-dimensional periodic profile is added to the surface to form a grating. If a periodic profile is introduced on the interface between the two media, a modification of the behaviors of the light refracted and reflected from the surface occurs. Specifically, a periodic grating is a type of one-dimensional photonic crystal which is found to create a sequence of diffracted rays from an incident light scattered by the grating. As is shown in the following, the nature of these diffracted rays is closely related to the wavelength of the light and the periodicity of the surface features.

In figure 3.6 a schematic drawing is shown for the scattering of an incident plane wave of light from a periodic one-dimensional grating. The grating is composed of a periodic interface which has a smallest repeat distance a along its mean plane. As in the case of the planar interface, the periodic surface separates an upper medium characterized by the material parameters $\varepsilon', \mu' > 0$ and a lower medium characterized by the parameters $\varepsilon,\mu > 0$. In the discussions to follow the incident plane wave travels in the medium below the scattering surface towards the interface. The incident wave is scattered at the grating interface into a set of diffracted waves propagating away from the interface within the medium below the surface [17].

The periodic scattering features of the grating are considered to weakly interact with the incident light so that the elements of the scattering transition matrix are treated within the Born approximation. For simplicity it is also assumed that the incident plane of the light is perpendicular to the grooves of the grating, with an electric field component parallel to the grooves of the grating. The weak scattering at the surface is then characterized by an effective scattering potential

$$v(\vec{r}) = \sum_n v_n e^{i\frac{2\pi}{a}nx}\delta(z) \tag{3.100}$$

where $\vec{r} = x\hat{i} + y\hat{j} + z\hat{k}$, $v_n = v_{-n}$ so that $v(\vec{r})$ is real, and n runs over the integers. From (3.100) it is seen that the mean plane of the scattering surface is the x–y plane,

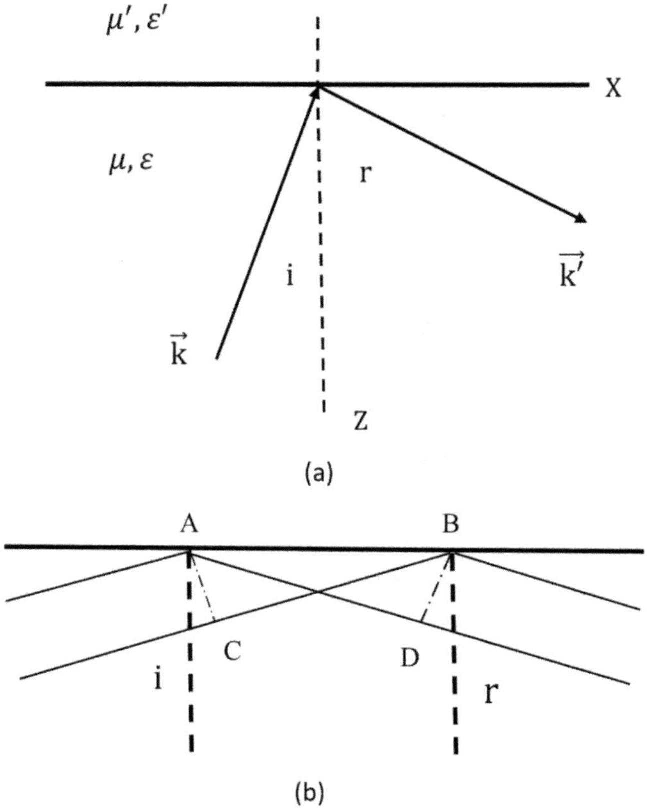

Figure 3.6. (a) Diffraction geometry for a grating in the x–y plane separating a medium μ,ε containing an incident and diffracted ray from a medium μ', ε'. The incident angle i is measured in the clockwise sense from the negative z-axis and the diffraction angle r is measured in the anti-clockwise sense from the negative z-axis. (b) The solid horizontal line is the grating surface and the dot–dashed segment AC is the wave front of the incident rays which make an incident angle i with the normal to the grating surface (dashed line). The dot–dashed segment BD is the wave front of the diffracted rays which make a diffraction angle r with the normal to the grating surface (dashed line). The length AB is the shortest repeat distance a of the grating ruled on the horizontal interface.

and the scattering potential is periodic in the x-direction with a smallest repeat distance a. The grating has total translational invariance along the y-direction.

Under these conditions the transition matrix for the scattering of an incident wave $|\psi_i\rangle \propto e^{i\vec{k}\cdot\vec{r}}$ to a scattered wave $|\psi_f\rangle \propto e^{i\vec{k}'\cdot\vec{r}}$ with both $\vec{k}$ and $\vec{k}'$ in the x–z plane is given by

$$\langle\psi_f \mid v(\vec{r}) \mid \psi_i\rangle \propto \int d^3r e^{-i\vec{k}'\cdot\vec{r}} v(\vec{r}) e^{i\vec{k}\cdot\vec{r}} \propto \sum_n v_n \delta_{k_x'-k_x,\frac{2\pi}{a}n}. \tag{3.101}$$

Consequently, if $\vec{k} = k_x\hat{i} + k_z\hat{k}$, then from the far right-hand side of (3.101)

$$\vec{k}' = \vec{k} + \frac{2\pi}{a}n\hat{i}. \tag{3.102}$$

For the energy of the light to be conserved in the light backscattered into the lower region in figure 3.6, it is required that the lengths of $\vec{k}$ and $\vec{k}'$ are the same. This requirement is set by the condition that

$$k_x = \frac{2\pi}{\lambda}\sin i = \frac{\pi}{a}n \qquad (3.103a)$$

or written in terms of the angle of incidence i

$$\sin i = n\frac{\lambda}{2a}. \qquad (3.103b)$$

Equation (3.103b) is recognized as the standard Bragg formula for the condition of diffraction.

An alternative statement of Bragg's laws in (3.103b) can also be obtained that gives a useful geometric interpretation in terms of both the incident and diffracted rays. Specifically, from (3.102) and the condition of energy conservation it follows that $k_x' = k_x + \frac{2\pi}{a}n$ where $k_x' = \frac{2\pi}{\lambda}\sin r$ and $k_x = \frac{2\pi}{\lambda}\sin i$ represent the behaviors of the x-components of the incident and diffracted rays. In terms of these components, the diffraction condition in (3.103) can now be reformulated and expressed in the form [17]

$$\sin r - \sin i = n\frac{\lambda}{a}. \qquad (3.104)$$

The relationship in (3.104) has a simple geometric interpretation.

3.4.1.2 Geometric interpretation of (3.104)

An interesting geometric interpretation of (3.104) is now given for the diffraction within the region below the grating surface. In this regard, consider the schematic in figure 3.6(b). The relationship in (3.104) follows directly from the ray geometry in figure 3.6(b). Specifically, treat the scattering at the grating for two points A and B that are separated by the grating periodicity a. In this geometry the difference in length of the path of the ray scattered at point A and the ray scattered at point B is given by [17]

$$\overline{AD} - \overline{CB} = a\left[\cos\left(\frac{\pi}{2} - r\right) - \cos\left(\frac{\pi}{2} - i\right)\right] = a[\sin r - \sin i]. \qquad (3.105)$$

For a phase coherent addition of these two rays, their path difference must be an integer multiple of the wavelength λ of the elastically scattered light. As a consequence of this (3.104) directly follows.

A similar treatment to that of the diffraction below the interface can also be made for the diffracted rays created by the scattering into the medium above the interface. This is left to the reader. We now discuss some of the interesting consequences of the diffraction conditions in (3.103) and (3.105).

3.4.1.3 Consequences of the diffraction conditions

From either (3.103) or (3.104) it is seen that the periodic surface not only allows for the presence of a reflected wave (i.e., $i = r$ in the case of (3.104) in which $n = 0$) but also from the form in (3.103b) allows for a discrete set of diffracted rays in which the condition $\left| n\frac{\lambda}{2a} \right| < 1$ is satisfied. In this regard, the smaller the ratio $\frac{\lambda}{2a}$ the more orders of diffraction exist in the system. In this scheme a diffracted ray with index n is referred to as a diffracted ray of order n.

Another interesting case of the conditions in (3.103) and (3.104) is known as the Littrow's configuration. This is the particular angle of incidence in which one of the diffractive orders travels antiparallel to the original incident light. From (3.104) the angle of incident i_L in this arrangement is given by [17]

$$\sin i_L = -\frac{n_L}{2}\frac{\lambda}{a} \tag{3.106}$$

where n_L is a negative integer. It is seen that, for a given ratio $\frac{\lambda}{a}$, there may be multiple solutions of equation (3.106). In this configuration the backscattered radiation is often referred to as a blaze.

Similar to the case of total internal reflection at a planar interface, there can also exist diffractive orders with angles of diffraction along the interface. As in the case of total internal reflection the rays diffracted along the interface are evanescent in the medium above the interface. The condition for this type of behavior is obtained from (3.104) rewritten in the form [17]

$$\sin r = n\frac{\lambda}{a} + \sin i. \tag{3.107}$$

Specifically, for $\left| n\frac{\lambda}{a} + \sin i \right| \geqslant 1$ the diffractive wave of order n becomes an evanescent wave similar to that of the solution along the planar interface in the case of total internal reflection. These solutions, however, do not form true surface waves such as those that are now discussed.

In addition to the scattering solutions at the planar surface, there can exist a range of surface electromagnetic waves. These excitations, which are bound to the surface, are now briefly discussed.

3.4.2 Theory of surface waves on periodic surfaces

Periodic surfaces can also support a variety of surface electromagnetic modes which are bound to and travel along the interface between two different media [18–22]. These are known as surface polaritons, surface plasmons, or surface plasmon–polaritons depending on the types of materials involved in their supporting interfaces [18–22]. Surface modes are independent from the scattering modes present at the interface and have amplitudes that decay exponentially with separation from the surface. In order for them to exist at the interface the surface waves require a particular set of conditions on the materials forming the surface and the geometry characterizing the surface. In the following, first the surface waves on a planar interface are reviewed, and this is followed by the brief discussion of the surface waves bound to a periodic interface.

For the treatment of surface electromagnetic waves on a planar interface, consider a surface located in the y–z plane and separating two isotropic homogeneous media. Above the surface is a medium of dielectric function $\varepsilon_>(\omega)$ and permeability $\mu_>(\omega) = 1$, and below the surface is a medium of dielectric function $\varepsilon_<(\omega)$ and permeability $\mu_<(\omega) = 1$. In the following considerations it is assumed that both media are linear optical media.

For this arrangement of dielectrics the wave equations obtained from the Maxwell equations in (3.93) are of the form [18]

$$\left[\nabla^2 + \frac{\varepsilon_i\omega^2}{c^2}\right]\vec{E}_i = 0 \tag{3.108}$$

$$\left[\nabla^2 + \frac{\varepsilon_i\omega^2}{c^2}\right]\vec{B}_i = 0$$

where $i = >,<$ indicate the regions above and below the surfaces. The boundary conditions at the interface of the two media require that the normal components of the vectors $\vec{D}$ and $\vec{B}$ are continuous and the tangential components of the vectors $\vec{E}$ and $\vec{H}$ are also continuous.

The solutions for surface waves propagating along the interface with an exponentially decaying amplitude in terms of the separation from the interface are found to be of the general form [18]

$$\vec{E}_>(x, y, z) = \left(\frac{ik}{\alpha_>(\omega)}, 0,1\right)E_>\exp[-\alpha_>(\omega)x + i(kz - \omega t)] \tag{3.109a}$$

above the surface, and

$$\vec{E}_<(x, y, z) = \left(-\frac{ik}{\alpha_<(\omega)}, 0,1\right)E_<\exp[\alpha_<(\omega)x + i(kz - \omega t)] \tag{3.109b}$$

below the surface where $\alpha_>(\omega)$, $\alpha_<(\omega) > 0$. (Notice that in these expressions, without a loss of generality, the surface wave is taken to propagate in the z-direction.) Here

$$\alpha_>^2(\omega) = k^2 - \frac{\omega^2}{c^2}\varepsilon_>(\omega) \tag{3.110a}$$

$$\alpha_<^2(\omega) = k^2 - \frac{\omega^2}{c^2}\varepsilon_<(\omega) \tag{3.110b}$$

and, from matching of the field boundary conditions on the $x = 0$ interface, it follows that [18]

$$E_> = E_< \tag{3.111}$$

$$\frac{\varepsilon_>(\omega)}{\varepsilon_<(\omega)} = -\frac{\alpha_>(\omega)}{\alpha_<(\omega)}. \tag{3.112}$$

In regard to the boundary conditions it is noted that the vectors $\vec{D}$, $\vec{B}$, and $\vec{H}$ are generated from $\vec{E}$ in (3.109) by an application of the Maxwell equations in (3.93).

From (3.112) it is found that, since both $\alpha_>(\omega)$ and $\alpha_<(\omega)$ are positive, the values of $\varepsilon_>(\omega)$ and $\varepsilon_<(\omega)$ must be of opposite sign in order to support a surface wave. This is a consequence of the evanescent character of the wave functions in the two different media. In addition, it is important to note that, unlike in the case of the propagation of bulk waves, the surface wave has both transverse and longitudinal field components.

As a specific example of the surface wave formulae, for the solution presented in (3.108) through (3.112), consider the case of a plasmon–polariton supported on a vacuum–metal interface. In the region above the interface, characterize the medium by $\varepsilon_>(\omega)$ of a metal described by the Drude form

$$\varepsilon_>(\omega) = \varepsilon_\infty\left(1 - \frac{\omega_p^2}{\omega^2}\right) \tag{3.113}$$

where ω_p is the plasma frequency of the metal. Below the interface, $\varepsilon_<(\omega) = 1$ characterizes the region of vacuum. From (3.112) the dispersion relation of the surface plasmon modes on the interface is given by [18]

$$\omega_{sp}^2 = \frac{1}{2\varepsilon_\infty}\left\{(1+\varepsilon_\infty)c^2k^2 + \varepsilon_\infty\omega_p^2 - \sqrt{\left((1+\varepsilon_\infty)c^2k^2 + \varepsilon_\infty\omega_p^2\right)^2 - 4\varepsilon_\infty^2c^2k^2\omega_p^2}\right\}. \tag{3.114}$$

Note that for a fixed k the surface plasmon mode frequency ω_{sp} is less than the frequency of light in vacuum, and it is also bounded from above by the limiting form $\omega_{sp}(k \to \infty) \to \frac{\omega_p}{\sqrt{1+\frac{1}{\varepsilon_\infty}}}$ [18].

As an example of the dispersion relation in (3.114) in figure 3.7 a plot is presented of $\frac{\omega_{sp}}{\omega_p}$ versus $\frac{ck}{\omega_p}$ for the case in which $\varepsilon_\infty = 1$. The plot exhibits a behavior which is typical of the general behavior of plasmon–polaritons for a medium characterized by $\varepsilon_>(\omega)$ of the form in (3.113). The dispersion relation is seen to be an increasing function of k, and at small wavelengths the frequency approaches the light line $\omega = ck$ which has been indicated in the plot. At large wave vectors the system ω_{sp} approaches a constant limiting form.

3.4.2.1 Surface electromagnetic waves on a periodic surface

Next a brief consideration is given of the much more difficult problem involving surface electromagnetic waves in the presence of a periodic surface profile on the interface between the two isotropic homogeneous linear dielectric media [19]. The introduction of the periodic profile opens up a band structure in the surface wave dispersion relation with pass and stop bands similar to those found in the bulk electromagnetic modes of three- and two-dimensional photonic crystals. In addition,

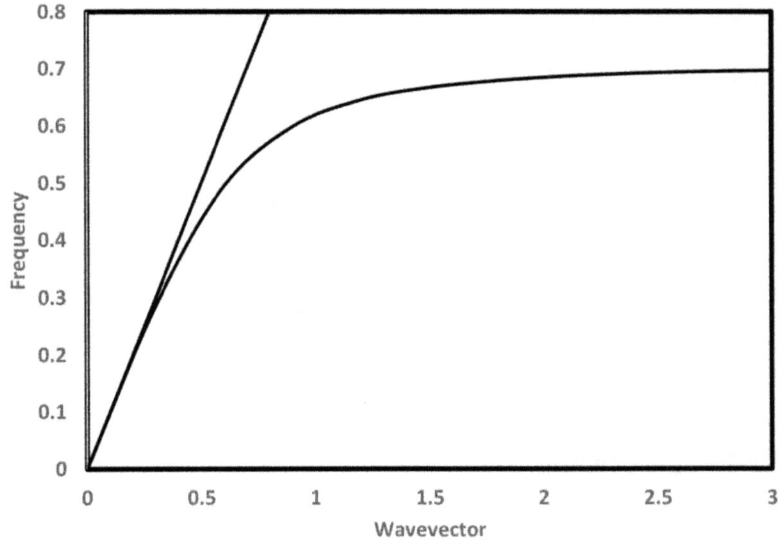

Figure 3.7. Plot of the surface plasmon–polariton dispersion in the normalized frequency, $\frac{\omega sp}{\omega p}$, versus the normalized wavenumber, $\frac{ck}{\omega p}$.

however, new regions also open in the frequency–wave vector space in which surface excitations are no longer bound to the surface. These regions form unstable modes, known as leaky modes, which originate on the surface but due to their instability radiate from the interface [19, 22].

For the following considerations the surface between the two media is described by the set of coordinates (x_s, y, z) where the x-coordinate of the surface is given by a periodic surface profiled function of the form [19]

$$x_s = \xi(z) \tag{3.115}$$

with a smallest repeat distance a along the x-axis.

Above the surface is a medium of dielectric function $\varepsilon_>(\omega)$ given in (3.113) and permeability $\mu_>(\omega) = 1$, and below the surface is a medium of dielectric function $\varepsilon_<(\omega) = 1$ and permeability $\mu_<(\omega) = 1$. As in the earlier surface wave considerations, both media are linear optical media. Specifically, the system is a direct modification of that of the flat interface problem.

The problem now is not as straightforward as that of the planar surface because the periodicity of the surface requires a solution of the general form [19]

$$\vec{E}(x, z) = \exp[i(kz - \omega t)]\vec{U}(x, z) \tag{3.116}$$

where $\vec{U}(x,z)$ is periodic in z with a smallest repeat distance a and decays exponential to zero as $x \to \pm\infty$. This complicates the application of the electromagnetic boundary conditions at the now periodic interface.

In addition to the wave function restrictions, there are also restrictions on the surface wave dispersion relations. Specifically, it is important to realize that, for a fixed wave vector k, surface wave modes cannot have frequencies greater than those of bulk modes of light $\omega = ck$. The reason for this is that if this were the case, the surface waves would decay into lower frequency bulk modes of the system. This frequency restriction represents another fundamental limitation on the existence of surface waves bound to the grating.

The above system has been worked out in the context of a lowest order perturbation treatment of the surface roughness given in (3.115) [19], and surface waves are found on the periodic interface subject to the earlier mentioned restrictions. The detailed treatment of the problem requires an extensive analysis so that for the details of the considerations the reader is referred to [19] and [22]. Later we shall return to the discussion of the related problem of surface electromagnetic waves in the context of rough surfaces. This will be considered in much greater detail.

3.5 Fiber Bragg gratings

In the context of reduced dimensional systems, another important application of photonic crystal technology occurs in fiber optic applications [9, 20, 21]. Optical fibers are essentially one-dimensional dielectric fibers which transport light in a manner reminiscent of the transport of electrical current by metallic wires. The fibers, however, are optical waveguides which transport energy ballistically rather than diffusively as in the case of electrical wires.

As an illustration, figure 3.8(a) shows a section of a cylindrical optical waveguide which is designed to carry optical energy along the direction of its cylinder axis. The light traveling in the fiber is bound to the fiber by its dielectric properties and their contrast with those of the world outside of the fiber. In this regard, the fiber often has a cylindrical variation of its dielectric properties as a function of distance from the cylinder axis. This variation is designed to increase the efficiency of the fiber in binding and delivering electromagnetic energy. The efficiency of the fiber in transmitting light of a given frequency then depends on the index of refraction of the materials from which it is composted and their geometric arrangement.

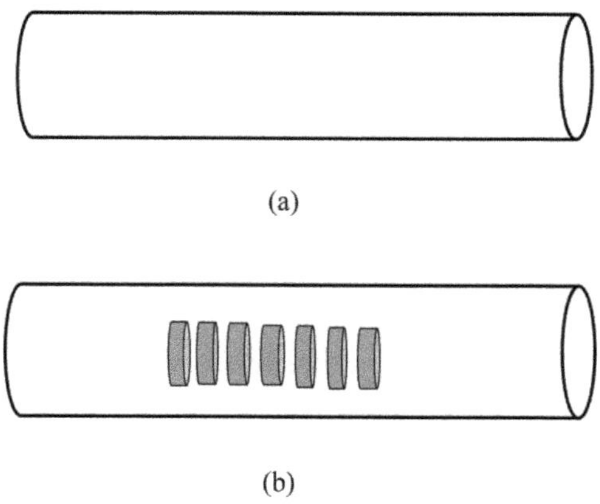

(a)

(b)

Figure 3.8. Fiber optics waveguide: (a) optical fiber without a grating, (b) optical fiber with an internal Bragg grating.

One application of the ideas of photonic crystals arises from using them to create an optical fiber with a specific variation of radial dielectric properties meant to facilitate the binding of electromagnetic modes to the fiber axis. This can be done by creating an appropriate modulation of the effective dielectric properties about the fiber axis. The photonic crystal operates here as a composite medium, and its band structure is of secondary importance. Another photonic crystal application is to create a periodic modulation in the radial properties of the fiber with stop and pass band sequences designed to help bind the modes of the fiber to its axis.

3.5.1 Cylindrically symmetric optical fibers

In figure 3.8(a) is a schematic representation of an infinitely long cylindrical waveguide. The fiber has complete translational symmetry along the horizontal axis through the center of the waveguide which is taken to be the z-axis. Outside the fiber is vacuum, and the dielectric properties of the fiber depend only on the distance from the axis of the fiber.

As a consequence of the translational symmetry parallel to the fiber axis, the modes bound to and propagating along the fiber axis have plane wave forms

$$e^{\pm ikz}\vec{U}(r,\,\theta) \tag{3.117}$$

which depends on the z-coordinate. (The dependence of the modes on the other cylinder coordinates r,θ in the plane perpendicular to the waveguide axis are not important here.) In this regard, for the following discussions of motion along the cylinder axis a single mode waveguide is treated, and only the forms $e^{\pm ikz}$ of its axially propagating modes are considered.

In (3.117), $k = \frac{2\pi}{\lambda}$ is the wave vector component in the z-direction where λ is the wavelength of the light moving parallel to the fiber axis. The values of k and λ depend on the geometry and the dielectric composition of the cylinder waveguide in figure 3.8 along which the modes propagate.

An interesting filtering mechanism based on the introduction of periodicity to the system can now be introduced. Unlike the earlier discussions of photonic crystal periodicity designed to affect the radial variation of the fiber dielectric properties and to aid the modal confinement, a useful filtering control occurs with the introduction of periodicity along the fiber axis. This is now discussed.

3.5.2 Periodicity along the fiber axis: Bragg gratings

By introducing the ideas of periodic dielectric properties into the design of optical fibers, the characteristic features of photonic crystals have found useful applications in fiber optics technology as filtering mechanisms. This introduction is done along the fiber axis as shown schematically in figure 3.8(b). Here a region of a finite array of periodic dielectric media is placed into a segment of an optical fiber which otherwise has translational symmetry along its fiber axis. Such a small array is found to introduce frequency pass and stop band regions into the fiber spectrum of transmitted light. These frequency pass and stop bands arise from Bragg interference effects, and the particular fiber design is known as a fiber Bragg grating. It provides a means of filtering out the transmission of certain frequency components of light from passing down the waveguide channel as the periodic dielectric variation along the fiber axis reflects them back towards their source.

To understand this, consider the fiber geometry in figure 3.8(b) which contains a small finite periodic array of dielectric disks along the fiber axis. Let the smallest repeat distance in the array be denoted a. A mode with a z-dependence of the form e^{ikz} will be scattered into modes of e^{ikz} and e^{-ikz} at each of the disks of the array. As a result, constructive interference will occur between the scattering of each of the disks preferentially into the e^{-ikz} modes when [20, 21]

$$\lambda = 2a \tag{3.118}$$

where $k = \frac{2\pi}{\lambda}$. This is the Bragg condition of phase constructive interference resulting in a back reflection into the fiber.

The wave vector of the light propagating in the waveguide along the z-direction, however, is related to the frequency of the light ω by

$$k = n_{eff}\frac{\omega}{c} = n_{eff}\frac{2\pi}{\lambda_0} \tag{3.119}$$

where n_{eff} is an effective dielectric for the motion along the axis of the fiber, and λ_0 is the wavelength of the light in vacuum. Here n_{eff} takes into account the fiber geometry and its dielectric properties so that it is influenced by both geometry as well as material properties.

Consequently, from (3.118) and (3.119) the Bragg condition for the reflection of the incident mode into a mode propagating opposite it is [20, 21]

$$\lambda_0 = 2n_{eff}a. \tag{3.119}$$

It is essentially the statement that the wavelength of the light which constructively interferes depends proportionally on the grating period a.

Light satisfying the Bragg condition in (3.119) is back reflected by the fiber grating. This allows for the properties of a one-dimensional photonic crystal to be used to filter the properties of the light passed down the waveguide, removing some of the frequency components from the light passed. Since the grating is not infinite in extent, however, the degree of the filtering effect and the range of the bandwidth that is filtered depends on the length of the grating placed within the optical fiber. By creating a number of such gratings within the fiber with different a, a series of different wavelengths can be filter from the fiber transmission spectrum [22].

References

[1] Joannopoulos J D, Vilenueve P R and Fan S 1995 *Photonic Crystals* (Princeton, NJ: Princeton University Press)

[2] Sakoda K 2001 *Optical Properties of Photonic Crystals* (Berlin: Springer)

[3] Favennec P N 2005 *Photonic Crystals: Towards Nanoscale Photonic* (Berlin: Devices Springer)

[4] McGurn A R 2002 *Survey of Semiconductor Physics Boer* (New York: Wiley) ch 13
[5] Arfken G B and Weber H J 2005 *Mathematical Methods for Physicists* 6th edn (Amsterdam: Elsevier)
[6] Marder M P 2010 *Condens. Matter Phys.* 2nd edn (Hoboken, NJ: Wiley)
[7] Maradudin A A and McGurn A R 1993 The photonic band structure of a truncated, two-dimensional, periodic medium *J. Opt. Soc. Am.* **10** 307
[8] McGurn A R 2005 Impurity mode techniques applied to the study of light sources *J. Phys.* **38** 2338
[9] Jackson J D 1999 *Classical Electrodynamics* 3rd edn (New York: Wiley)
[10] McGurn A R 1996 Green's-function theory for row and periodic defect arrays in photonic band structures *Phys. Rev.* B **53** 7059
[11] Galin J S 2000 *Impurity Scattering in Metallic Alloys* (New York: Springer)
[12] Taflove A 1995 *Computational Electrodynamics: The Finite-Difference Time-Domain Method* (Boston, MA: Artech House)
[13] El Misilmani H M, Kabalan K Y, Abou-Shahine M Y and Al-Husseini M 2015 A method of moment approach in solving boundary value problems *J. Electromagn. Anal. Appl.* **7** 61
[14] Gibson W C 2008 *The Method of Moments in Electromagnetics* (Boca Raton, FL: Chapman and Hall/CRC)
[15] Humphries S 1997 *Finite-Element Methods for Electromagnetics* (Boca Raton, FL: CRC)
[16] Mekis A, Fan S and Joannopoulos J D 1999 Absorbing boundary conditions for FDTD simulations of photonic crystal waveguides *IEEE Microw. Guid. Wave Lett.* **9** 502
[17] Chartier G 2005 *Introduction to Optics* (New York: Springer)
[18] Wallis R F 1986 *Electromagnetic Surface Excitations* ed R F Wallis and G I Stegeman (Berlin: Springer)
[19] Maradudin A A 1986 *Electromagnetic Surface Excitations* ed R F Wallis and G I Stegeman (Berlin: Springer)
[20] Wang C and Yao J 2015 *Microwave Photonics* ed C H Lee (Milton Park: Taylor and Francis)
[21] Hill K O and Meltz G 1997 Fiber Bragg grating technology fundamentals and overview *J. Lightwave Technol.* **15** 1263
[22] McGurn A R 2018 *Nanophotonics* (Cham: Springer)

IOP Publishing

Introduction to Nonlinear Optics of Photonic Crystals and Metamaterials (Second Edition)

Arthur R McGurn

Chapter 4

Basic properties of metamaterials

Metamaterials are engineered materials that include specific artificial mesoscopic features in their designs [1–5]. The mesoscopic features are introduced so as to tailor the material to exhibit specific permittivity and permeability properties when interacting with radiation of wavelengths greater than the mesoscopic inclusions. To such externally applied electromagnetic waves metamaterials appear homogeneous just as crystalline materials appear homogeneous for electromagnetic waves at optical wavelengths. The interest in the artificial mesoscopic inclusions is that they can be made to exhibit magnetic resonances at much higher frequencies of electromagnetic waves than those of the magnetic resonances of atoms and molecules of natural crystals [1]. This allows systems containing these features to exhibit high frequency regions of negative permeability, whereas atomic and molecular systems only display negative permeability at much lower frequencies than those of the metamaterials. The high frequency magnetic resonances when combined with a background dielectric medium of negative permittivity gives rise to metamaterials exhibiting a new property of negative refractive indices. Regions of frequencies with simultaneous negative permittivity and permeability have not been observed in naturally occurring materials and are only now available in designs of metamaterials and for certain configurations of photonic crystals. As a result the possibility of negative refractive index materials extends what was once considered a fundamental limitation on optical design. We shall see that the greater refractive properties found in materials with negative refractive indices extends refractive angles over a greater range of values.

In the following the properties of the basic split ring resonator (SRR) units forming metamaterials are explained along with how they can be arrayed in three dimensions to form bulk materials with engineered diamagnetic response properties. This is followed by a discussion of the properties of negative refractive index materials.

doi:10.1088/978-0-7503-3579-9ch4

4.1 Properties of SRRs and SRR arrays

The basic design of the inclusions engineered into metamaterials is an SRR [1–4]. There are many variations on the design that are made for engineering considerations, and the reader is referred to the literature and some remarks in the chapter 9 of this book for more details. In its basic design the SRR is a ring with a gap cut into it (see figure 4.1). The ring acts as an inductance and the gap acts as a capacitor so that the structure is an LC resonant circuit. The natural frequency of the SRR is then given by

$$\omega_0 = \frac{1}{\sqrt{LC}} \tag{4.1}$$

where L and C are the self-inductance and capacitance of the split ring, and this is the resonant frequency of the interaction of the SRR with an external driving electromagnetic wave. In addition to the self-inductance the neighboring rings in the metamaterial have a mutual inductance.

When an SRR is driven by an external electromagnetic wave of frequency, ω, as the wave frequency is passed through the natural frequency ω_0 of the SRR the SRR undergoes a resonant interaction with the wave [6–8]. To understand the operation of an SRR interacting with an external frequency-dependent wave consider the planar loop in figure 4.2(a) containing an inductor and a capacitor. From Faraday's law a time-dependent magnetic field perpendicular to the plane of the loop induces an electromotive force (emf) in the loop. In the figure the magnetic field is perpendicular to the page with the positive sense of the field out of the page, and the positive current in the loop is in the anti-clockwise direction. In the presence of the induced emf from the time-dependent field the circuit is essentially a forced LRC electromagnetic oscillator of the type shown in figure 4.2(b). The loop acts both as

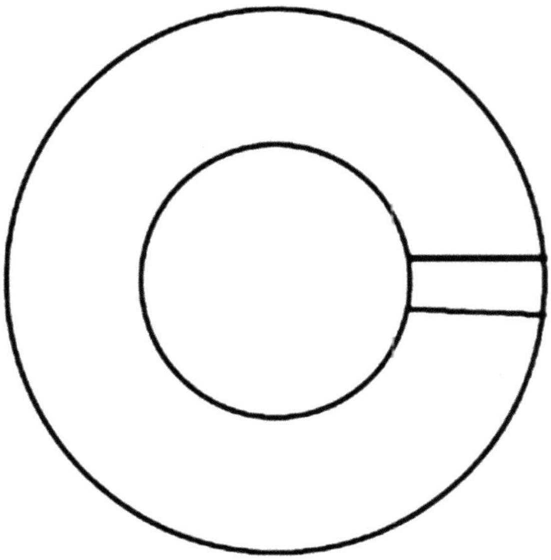

Figure 4.1. Basic design of an idealized SRR unit, a metallic ring with a gap filled with dielectric material.

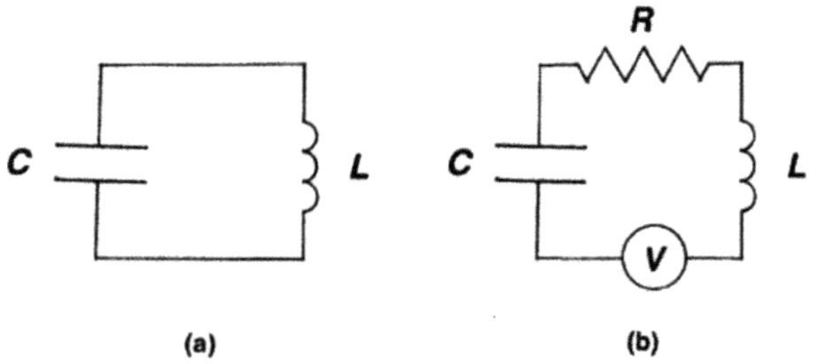

Figure 4.2. Schematics of: (a) an *LC* ring with a time varying magnetic field perpendicular to the plane of the ring, and (b) an *LRC* forced harmonic oscillator circuit.

the self-inductance of the circuit and the origin of the driving emf from its Faraday's law interaction with the external field. The equation of the driven oscillator is

$$L\frac{d^2Q}{dt^2} + R\frac{dQ}{dt} + \frac{Q}{C} = V \tag{4.2}$$

where Q is the charge on the capacitor and V is the forcing emf from Faraday's law.

For a magnetic field of the form $B(t) = B_0 \cos\omega t$ the forcing emf in the *LRC* circuit is

$$V = V_0 \sin\omega t \tag{4.3}$$

where $V_0 = \omega A B_0$ for a loop of area A. From the theory of the *LRC* forced oscillator the response of the oscillator to the driving emf is given by the impendence of the circuit

$$Z(\omega) = \sqrt{R^2 + \left(\omega L - \frac{1}{\omega C}\right)^2} \tag{4.4}$$

relating the induced current in the circuit to the induced emf and the phase difference, φ, between the current and the driving emf

$$\varphi = \tan^{-1}\left(\frac{\omega L - \frac{1}{\omega C}}{R}\right). \tag{4.5}$$

Consequently, the current in the loop is given by

$$I(t) = I_0 \sin(\omega t - \varphi) \tag{4.6}$$

where

$$I_0 = \frac{V_0}{Z(\omega)}. \tag{4.7}$$

It is interesting to note that at the resonant condition $\omega = \omega_0 = \frac{1}{\sqrt{LC}}$ so that from (4.5) $\varphi = 0$, and the induced current and emf are in phase. Consequently, the time average resistive power losses in the circuit at resonance, $P_{avg} = \left(\frac{I_0}{\sqrt{2}}\right)^2 R = \left(\frac{V_0}{\sqrt{2}}\right)^2 \frac{R}{Z^2}$, are a maximum. In addition, the magnetic moment of the loop, μ, is given by

$$\mu(t) = I(t)A \tag{4.8}$$

and the potential energy of the interaction of the field with the dipole moment is

$$U(t) = -\mu(t)B(t). \tag{4.9}$$

Substituting in (4.9) gives

$$U(t) = -AI_0B_0 \cos(\omega t) \sin(\omega t - \varphi) \tag{4.10}$$

which on time average gives

$$\bar{U} = \frac{AI_0B_0}{2}\sin(\varphi). \tag{4.11}$$

From (4.11) it is seen that another consequence of the absence of a phase difference between the induced current and emf is that at resonance $U = 0$ and the effective magnetic moment of the loop is zero.

The regions of interest for metamaterial applications are at frequencies slightly above or below the resonant frequency. Expanding the frequency of the electromagnetic wave about the resonant frequency of the SRR

$$\omega = \omega_0 + \Delta\omega \tag{4.12}$$

gives from (4.5) the phase in terms of $\Delta\omega$

$$\varphi \approx 2\frac{L}{R}\Delta\omega \tag{4.13}$$

and from (4.4) the impedance

$$Z \approx R\sqrt{1+4\left(\frac{L}{R}\right)^2(\Delta\omega)^2} \tag{4.14}$$

in terms of $\Delta\omega$. From (4.13) and (4.14) the average power is given by

$$P_{avg} = \left(\frac{V_0}{\sqrt{2}}\right)^2 \frac{R}{Z^2} \approx \left(\frac{V_0}{\sqrt{2}}\right)^2 \frac{1}{R\left(1+4\left(\frac{L}{R}\right)^2(\Delta\omega)^2\right)} \tag{4.15}$$

so that for $\Delta\omega \neq 0$ the average power in (4.15) is less than at the $Z = R$ resonance condition.

In addition, from (4.11), (4.13), and (4.14) the average potential energy for the interaction with the magnetic field is

$$\bar{U} \approx AI_0B_0\frac{L}{R}\Delta\omega. \tag{4.16}$$

The sign of the average potential energy then depends on the sign of $\Delta\omega$. For positive $\Delta\omega$ the loop is diamagnetic and for negative $\Delta\omega$ the loop is paramagnetic. Defining the effective magnetic moment of the loop as

$$\mu_{\text{eff}} = -\frac{\bar{U}}{B_0} \tag{4.17}$$

gives the effective moment

$$\mu_{\text{eff}} = -\frac{L}{R^2}\frac{\omega A^2 B_0}{\sqrt{1+4\left(\frac{L}{R}\right)^2(\Delta\omega)^2}}\Delta\omega. \tag{4.18}$$

Near resonance the basic properties of the energy loss and effective magnetic moment of the SRR are simply related to the difference between the frequency of the electromagnetic wave and the SRR resonant frequency [2]. Plots of these properties are presented in figure 4.3. From (4.15) and figure 4.3(*a*) it is seen that the power loss

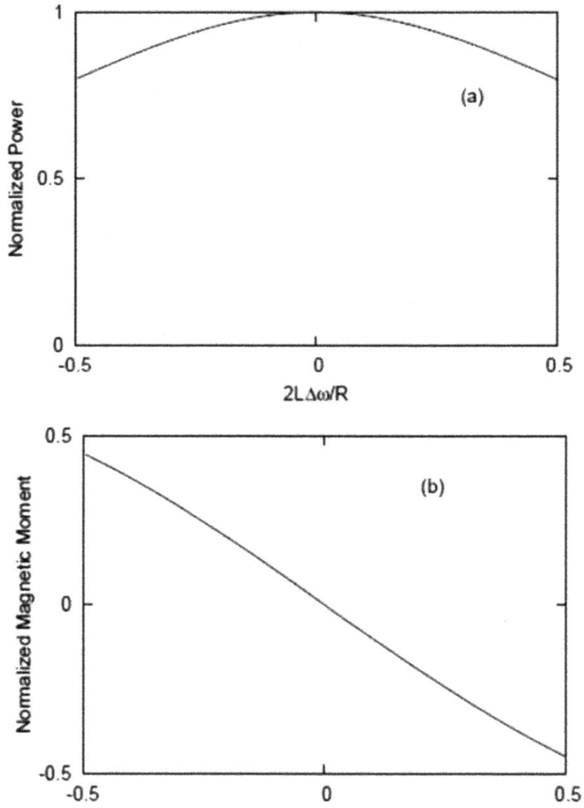

Figure 4.3. Power loss of the field to the SRR and the effective magnetic moment of the SRR: (a) the normalized power, $2RP_{\text{avg}}/V_0^2$, and (b) the normalized magnetic moment, $2R\mu_{\text{eff}}/\omega A^2 B_0$, both plotted versus $2L\ \Delta\ \omega/R$ where $\Delta\omega = 0$ at resonance.

in the *LRC* circuit exhibits a maximum at the $\Delta\omega = 0$ resonance condition while from (4.18) and its plot in figure 4.3(*b*) the effective magnetic moment passes through a sign change. By setting the values of $\frac{L\omega A^2 B_0}{R^2}$ for the amplitude and $\frac{L}{R}$ for the width of the resonance of the effective magnetic moment, the SRR passes through regions of enhanced paramagnetic and diamagnetic responses as the frequency of the electromagnetic wave passes through the SRR resonance. The region of enhanced diamagnetic response has been of great recent interest as in this region it is possible to use SRRs to facilitate the design of materials with negative permeability. In the additional presence of negative permittivity a negative refractive index is exhibited [3].

The results in figure 4.3 also illustrate a problem with the implementation of SRR metamaterials for their negative refractive index possibilities. The maximum of the losses in the materials occurs near the region of enhanced diamagnetism so that the region of losses can extend into the region of negative refractive index. It is one of the design problems of metamaterials to find ways of lessening the losses of the materials while obtaining a good negative index of refraction. The resonance nature of the negative index presents another design problem in that resonances are typically associated with instabilities of the system at a single frequency. This tends to limit the negative index effect to single frequencies, and the applications which are discussed later have predominantly been studied for narrow frequency bands.

For a material to display a negative index of refraction it must simultaneously have a negative permittivity and permeability [4, 5]. From the discussions above it is seen that a single SRR can be tuned to exhibit a negative permeability for an electromagnetic wave with a magnetic field polarized perpendicular to the plane of the SRR. To design a bulk material displaying a three-dimensional, homogeneous, isotropic diamagnetic response for electromagnetic waves propagating in a general direction within the bulk it is necessary to make an array of SRRs. An example of such a three-dimensional array is obtained by placing SRRs periodically in the x–y, y–z, and x–y planes of the bulk medium. This system of SRRs then forms a three-dimensional crystal with SRRs as the basis. The resonance conditions of the SRRs of the array are then tuned so that they give an enhanced diamagnetic response for waves with long wavelength compared to the size of the SRRs. This creates a bulk material with a negative permeability. To design a material with a negative refractive index it only remains to infuse into the SRR crystal a background material of negative permittivity. We now turn to a discussion of the properties of negative refractive index materials.

4.2 Negative refractive index metamaterials

To understand the nature of negative refractive materials begin by reviewing the solution for an electromagnetic plane wave propagating in a general electromagnetic medium [4, 5]. Consider a plane wave moving along the x-axis with $\vec{E}$ and $\vec{B}$ polarized, respectively, along the y- and z-axes. From Faraday's law

$$\frac{\partial E_y}{\partial x} = -\frac{\partial B_z}{\partial t} \tag{4.19}$$

and from Ampere's law

$$-\frac{\partial B_z}{\partial x} = \mu\varepsilon\frac{\partial E_y}{\partial t} + \mu\sigma E_y \tag{4.20}$$

where ε is the permittivity, μ is the permeability, and σ is a small conductivity. The plane wave solutions of (4.19) and (4.20) are of the form

$$E_y = E_0 \mathrm{e}^{\mathrm{i}(kx-\omega t)} \tag{4.21a}$$

and

$$B_z = B_0 \mathrm{e}^{\mathrm{i}(kx-\omega t)}. \tag{4.21b}$$

Substituting (4.21) into (4.19) and (4.20) gives the matrix equation for the field amplitudes

$$\begin{vmatrix} k & -\omega \\ -(\mu\varepsilon\omega + i\mu\sigma) & k \end{vmatrix} \begin{vmatrix} E_0 \\ B_0 \end{vmatrix} = 0 \tag{4.22a}$$

and the dispersion relation

$$k^2 - (\mu\varepsilon\omega^2 + i\mu\sigma\omega) = 0. \tag{4.22b}$$

For fixed $\omega > 0$ expanding to first order in σ gives

$$k = \pm\sqrt{\mu\varepsilon}\,\omega \pm \frac{i\mu\sigma}{2\sqrt{\mu\varepsilon}} \tag{4.23}$$

where the upper signs come from taking the square root to be a positive number and the lower signs come from taking the square root to be a negative number. The positive sign gives a phase velocity in the positive x-direction while the negative sign gives a phase velocity in the negative x-direction.

Substituting (4.23) into (4.21) gives

$$E_y = E_0 \mathrm{e}^{\mathrm{i}(\pm\sqrt{\mu\varepsilon}\,\omega x-\omega t)} \mathrm{e}^{\mp\frac{\mu\sigma}{2\sqrt{\mu\varepsilon}}x} \tag{4.24a}$$

and

$$B_z = B_0 \mathrm{e}^{\mathrm{i}(\pm\sqrt{\mu\varepsilon}\,\omega x-\omega t)} \mathrm{e}^{\mp\frac{\mu\sigma}{2\sqrt{\mu\varepsilon}}x} \tag{4.24b}$$

where for the upper sign

$$\begin{vmatrix} E_0 \\ B_0 \end{vmatrix} = \begin{vmatrix} \frac{1}{\sqrt{\mu\varepsilon}} \\ 1 \end{vmatrix} A \tag{4.25a}$$

and for the lower sign

$$\begin{vmatrix} E_0 \\ B_0 \end{vmatrix} = \begin{vmatrix} -\frac{1}{\sqrt{\mu\varepsilon}} \\ 1 \end{vmatrix} A. \tag{4.25b}$$

Here A is the amplitude of the magnetic induction.

From (4.24a) and (4.24b) it is seen that in the $\sigma \to 0$ limit the Poynting vector is given by

$$\vec{S} = \frac{1}{2}\frac{1}{\mu}\vec{E} \times \vec{B}^* = \pm\frac{1}{2}\frac{1}{\mu}\frac{1}{\sqrt{\mu\varepsilon}} \quad A\,|^2\,\hat{\imath}. \tag{4.26}$$

From the form of the one-dimensional Poynting vector in (4.26) it is seen that the direction of the energy flow is determined by the sign of the permeability of the medium. In the case of a positive indexed material (i.e., for ε, $\mu > 0$) the flow of energy in the system is parallel to the wave vector of the plane wave where from (4.24) the wave vector is given by $\pm\,\mu\varepsilon\omega i$. In the case of a negative indexed material (i.e., for ε, $\mu < 0$) the flow of energy in the system is anti-parallel to the wave vector where the wave vector is again given by $\pm\,\mu\varepsilon\omega i$. Consequently, the signs of the one-dimensional wave vector and Poynting vector agree with one another in a positive indexed medium and are opposite one another in a negative index medium.

Including consideration of small non-zero conductivity in the electromagnetic solutions in (4.24) introduces a spatial decay term into the solutions. The spatial decay for both the electric field and the magnetic induction is given by $e^{\mp\frac{\mu\sigma}{2\sqrt{\mu\varepsilon}}x}$. In the case of a positive indexed material $\mu\sigma > 0$, and the amplitude of the waves decay in the direction of the energy flow. In addition, the energy flow is parallel to the direction of the wave vector. In the case of a negative indexed material $\mu\sigma < 0$, and the amplitude of the waves again decay in the direction of the energy flow. However, in this case the wave vector of the waves propagating in the negative indexed medium is now anti-parallel to the energy flow. The solutions for both the positive and negative indexed media make sense. In both cases they describe flows of energy which decay as the flow of energy advances through their respective media.

It may seem contradictory that for propagation in a negative refractive indexed medium the wave vector and Poynting vector are anti-parallel. This is not the case. To understand this, first consider a plane wave pulse of radiation propagating in a non-dissipative positive indexed medium. The pulse is localized in space and propagates as a spatially localized increase in energy that is moving in the direction of the wave vector. The wave vector and Poynting vector are parallel and the electric and magnetic energy density in the pulse are given, respectively, by

$$U_E = \frac{1}{2}\varepsilon\,|\,E(x,\,t)|^2 \tag{4.27a}$$

$$U_B = \frac{1}{2}\frac{1}{\mu}\,|\,B(x,\,t)\,|^2. \tag{4.27b}$$

In the absence of dispersion or for weak dispersion, the magnitude of the Poynting vector, giving the flow of energy in the direction of the wave vector, is related to the energies in (4.27) by

$$S = 2\,|\,U_E\,|\,v = 2\,|\,U_B\,|\,v \tag{4.28}$$

where v is the speed of light in the positive medium. Since ε, $\mu > 0$ the Poynting vector in (4.27) through (4.28) represents a net flow of positive energy carried by the pulse.

Now consider the flow of energy associated with a spatially localized field pulse moving in a negative indexed medium. Equations (4.27) and (4.28) also hold for such a pulse in the negative indexed medium. However, now $\varepsilon,\mu < 0$. The energy densities in (4.27) are negative and the pulse, written in the field amplitudes $|E(x, t)|$ and $|B(x, t)|$, shows up as a spatially localized decrease in energy. Though the pulse moves in a direction parallel to the wave vector, it is a pulse of energy decrease rather than a pulse of energy. In terms of the net energy in the system, a flow of negative energy past a point in space can be viewed as a flow of positive energy in the opposite direction. For the negative indexed medium, therefore, the amplitude of the Poynting vector in (4.28) represents a pulse of energy decrease traveling parallel to the wave vector or, equivalently, the flow of an energy pulse in the direction anti-parallel to the wave vector.

The flow of energy in positive and negative index media is similar to the flow of electrical current carried by the motion of holes and electrons in a semiconducting medium. In semiconductors, positive charged holes move parallel to the electric field to create the same electric current as negative charged electrons moving anti-parallel to the electric field. In the absence of an external magnetic field, the flow of the positive and negative charges cannot be distinguished. As shall be seen later, this analogy can be extended to considerations of the flow of light energy past an interface between a positive and negative index medium and the electrical current flow of electrons and holes through a p–n semiconductor junction. In these considerations a pulse of light energy is an energy particle in a positive indexed optical medium and a pulse of decrease of light energy is an energy hole in a negative indexed optical medium. The energy moves through the positive (negative) indexed medium by the motion of energy particles (energy holes), just as electrons carry the current in n-type semiconductors and holes carry the current in p-type semiconductors [9].

4.3 Refraction at the interface between positive and negative index media

As an example of the new properties introduced into optics by negative indexed materials, the refraction of light at the interface between positive and negative indexed media is considered [4, 5]. In a first case the refraction of light originating in the positive indexed medium is treated. This is followed by a treatment of the refraction of light originating in the negative indexed medium. Following these treatments a discussion is given of the design of a perfect lens.

4.3.1 Light originating in positive index media

Consider the planar interface between a positive and negative indexed medium. The positive indexed medium 1 (described by parameters μ_1, $\varepsilon_1 > 0$) is in the region $y > 0$, and the negative indexed medium 2 (described by parameters μ_2, $\varepsilon_2 < 0$) is in the region $y < 0$.

In terms of the angles defined in figure 4.4 the wave incident on the interface from medium 1 is given by

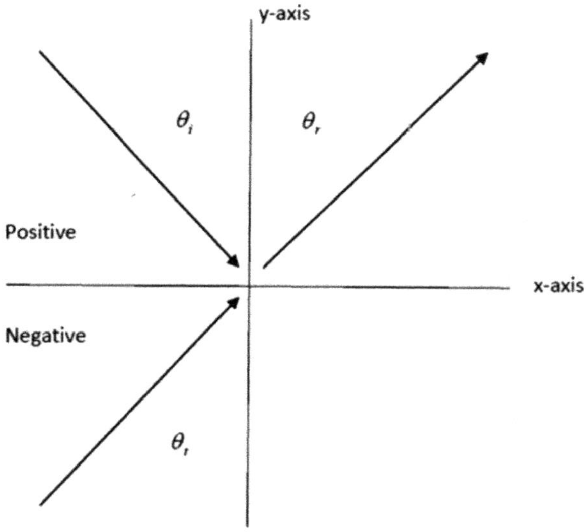

Figure 4.4. Refraction at a planar interface for incident light in a positive medium and refracted light in a negative medium. The arrows indicate the wave vectors and as represented in the figure θ_i, θ_r, $\theta_t > 0$.

$$\vec{E}_I = E_I(\cos\theta_i\hat{i} + \sin\theta_i\hat{j})e^{i[k(\sin\theta_i x - \cos\theta_i y) - \omega t]} \tag{4.29a}$$

$$\vec{B}_I = E_I\frac{k}{\omega}\hat{k}e^{i[k(\sin\theta_i x - \cos\theta_i y) - \omega t]} \tag{4.29b}$$

and the reflected wave in medium 1 is given by

$$\vec{E}_R = E_R(-\cos\theta_r\hat{i} + \sin\theta_r\hat{j})e^{i[k(\sin\theta_r x + \cos\theta_r y) - \omega t]} \tag{4.30a}$$

$$\vec{B}_R = E_R\frac{k}{\omega}\hat{k}e^{i[k(\sin\theta_r x + \cos\theta_r y) - \omega t]} \tag{4.30b}$$

where

$$k = \sqrt{\mu_1\varepsilon_1}\omega. \tag{4.31}$$

The wave vectors of the incident and reflected waves are

$$\vec{k}_i = k(\sin\theta_i, -\cos\theta_i, 0) \tag{4.32a}$$

$$\vec{k}_r = k(\sin\theta_r, \cos\theta_r, 0) \tag{4.32b}$$

and their respective Poynting vectors are

$$\vec{S}_I = \frac{1}{2}\frac{1}{\mu_1}\frac{1}{\omega} | E_I |^2 \vec{k}_i \tag{4.33a}$$

$$\vec{S}_R = \frac{1}{2}\frac{1}{\mu_1}\frac{1}{\omega} | E_R |^2 \vec{k}_r. \tag{4.33b}$$

Equations (4.32) and (4.33) show the standard result of energy flow into and out of the interface between the two media. Since $\mu_1 > 0$ the Poynting vectors of both the incident and reflected waves are parallel to their respective wave vectors.

For the case in which medium 2 is a negative indexed medium (i.e., ε_2, $\mu_2 < 0$) care must be taken to choose a solution in medium 2 representing an energy flow away from the interface. The correct form of the solution for the transmitted wave under these conditions is given by

$$\vec{E}_{\mathrm{T}} = E_{\mathrm{T}}(-\cos\theta_{\mathrm{t}}\hat{i} + \sin\theta_{\mathrm{t}}\hat{j})e^{i[q(\sin\theta_{\mathrm{t}}x+\cos\theta_{\mathrm{t}}y)-\omega t]} \tag{4.34a}$$

$$\vec{B}_{\mathrm{T}} = E_{\mathrm{T}}\frac{q}{\omega}\hat{k}e^{i[q(\sin\theta_{\mathrm{t}}x+\cos\theta_{\mathrm{t}}y)-\omega t]} \tag{4.34b}$$

where we use the angles as defined by figure 4.4 for the case of a negative medium 2. Specifically, the positive angle θ_{t} must be in the third quadrant and measured from the y-axis. The transmitted wave in equation (4.34) now has a wave vector and a Poynting vector of the respective forms

$$\vec{q}_{\mathrm{t}} = q(\sin\theta_{\mathrm{t}}, \cos\theta_{\mathrm{t}}, 0) \tag{4.35}$$

where $q = \sqrt{\mu_2\varepsilon_2}\,\omega$, and

$$\vec{S}_{\mathrm{T}} = \frac{1}{2}\frac{1}{\mu_2}\frac{1}{\omega} \mid E_{\mathrm{T}} \mid^2 \vec{q}_{\mathrm{t}} \tag{4.36}$$

which for $\mu_2 < 0$ represent an energy flow away from the interface and into the third quadrant.

For a comparison, consider the case in which medium 2 is positive indexed (i.e., ε_2, $\mu_2 > 0$) rather than negative indexed. Now the transmitted wave in medium 2 is given by

$$\vec{E}_{\mathrm{T}} = E_{\mathrm{T}}(\cos\theta_{\mathrm{t}}\hat{i} + \sin\theta_{\mathrm{t}}\hat{j})e^{i[q(\sin\theta_{\mathrm{t}}x-\cos\theta_{\mathrm{t}}y)-\omega t]} \tag{4.37a}$$

$$\vec{B}_{\mathrm{T}} = E_{\mathrm{T}}\frac{q}{\omega}\hat{k}e^{i[q(\sin\theta_{\mathrm{t}}x-\cos\theta_{\mathrm{t}}y)-\omega t]} \tag{4.37b}$$

where, in the case of a positive medium 2, the positive angle θ_{t} must be in the fourth quadrant and measured from the y-axis. The wave vector and Poynting vector of the transmitted wave, respectively, are given by

$$\vec{q}_{\mathrm{t}} = q(\sin\theta_{\mathrm{t}}, -\cos\theta_{\mathrm{t}}, 0) \tag{4.38}$$

$$\vec{S}_{\mathrm{T}} = \frac{1}{2}\frac{1}{\mu_2}\frac{1}{\omega} \mid E_{\mathrm{T}} \mid^2 \vec{q}_{\mathrm{t}} \tag{4.39}$$

where

$$q = \sqrt{\mu_2\varepsilon_2}\,\omega. \tag{4.40}$$

Since $\mu_2 > 0$ it is seen that the wave vector and Poynting vector are parallel, and the energy of the transmitted wave correctly flows away from the interface and into the fourth quadrant.

It is interesting to note the similarity of the solutions in (4.34) through (4.36) for the negative indexed medium 2 with those of the reflected wave for the positive indexed medium 1 given in (4.30), (4.32b), and (4.33b). In particular, both solutions have wave vectors with positive y-components. Due to the sign difference between $\mu_1 > 0$ and $\mu_2 < 0$, however, the two solutions represent energy flows away from the $y = 0$ plane in opposite directions along the y-axis. Specifically, it is seen from (3.35) and (4.36) that for $\mu_2 < 0$ the wave vector and Poynting vectors of the transmitted wave are anti-parallel. Though the wave vector has a positive component along the y-axis, from (4.36) the energy flow of the transmitted wave is in the negative y-direction, away from the interface. In the case of the reflected wave, however, the wave vector and Poynting vectors are parallel and the reflected wave has an energy flow along the positive y-axis.

In addition, it should be noted that, due to the translational symmetry of the planar interface at $y = 0$ between media 1 and 2, the solutions in (4.29), (4.30), (4.34), and (4.37) all have wave vectors with equal x-components. This requires that $\theta_i = \theta_r$ and $k\sin\theta_i = q\sin\theta_t$. These results arise solely from the translational symmetry and are independent of whether or not media 1 and 2 are positive or negative indexed materials.

To obtain the electric field amplitudes in these solution forms it is necessary to match the boundary conditions at the interface between the two media. From the continuity of the component of the electric field tangent to the interface it follows that

$$E_I - E_R = \alpha E_T \tag{4.41a}$$

where $\alpha = \frac{\cos\theta_t}{\cos\theta_i}$ when medium 2 has a positive index and $\alpha = -\frac{\cos\theta_t}{\cos\theta_i}$ when medium 2 has a negative index. From the continuity of the component of the magnetic field tangent to the interface it follows that

$$E_I + E_R = \beta E_T \tag{4.41b}$$

where $\beta = \frac{\mu_1\sqrt{\mu_2\varepsilon_2}}{\mu_2\sqrt{\mu_1\varepsilon_1}}$ when medium 2 is either negative or positive indexed. Solving these gives

$$E_T = \frac{2}{\alpha + \beta}E_I \tag{4.42a}$$

$$E_R = \frac{\beta - \alpha}{\alpha + \beta}E_I. \tag{4.42b}$$

4.3.2 Light originating in negative indexed media

Next, consider the planar interface between a negative indexed medium 1 (i.e., $\mu_1, \varepsilon_1 < 0$) in the region $y > 0$, and the positive indexed medium 2 (i.e., $\mu_2, \varepsilon_2 > 0$) is in the region $y < 0$. The incident wave is now in the negative indexed medium and the transmitted wave is in the positive indexed medium [4, 5].

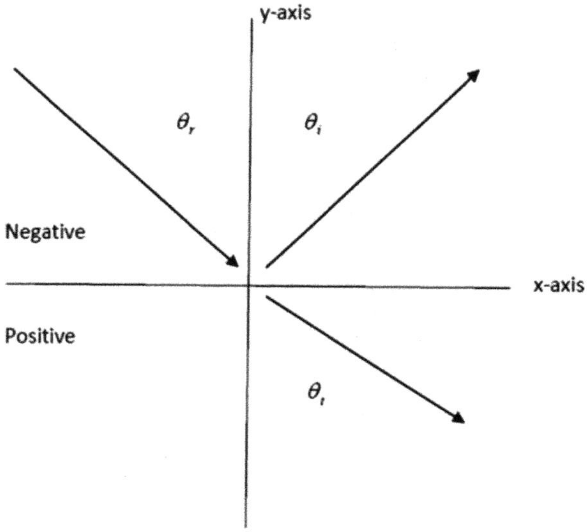

Figure 4.5. Refraction at a planar interface for incident light in a negative medium and refracted light in a positive medium. The arrows indicate the wave vectors and as represented in the figure θ_i, θ_r, $\theta_t > 0$.

In terms of the angles defined in figure 4.5 the wave incident on the interface from medium 1 is given by

$$\vec{E}_I = E_I(\cos\theta_i\hat{i} - \sin\theta_i\hat{j})e^{i[k(\sin\theta_i x+\cos\theta_i y)-\omega t]} \tag{4.43a}$$

$$\vec{B}_I = -E_I\frac{k}{\omega}\hat{k}e^{i[k(\sin\theta_i x+\cos\theta_i y)-\omega t]} \tag{4.43b}$$

and the reflected wave in medium 1 is given by

$$\vec{E}_R = E_R(\cos\theta_r\hat{i} + \sin\theta_r\hat{j})e^{i[k(\sin\theta_r x-\cos\theta_r y)-\omega t]} \tag{4.44a}$$

$$\vec{B}_R = E_R\frac{k}{\omega}\hat{k}e^{i[k(\sin\theta_r x-\cos\theta_r y)-\omega t]} \tag{4.44b}$$

where

$$k = \sqrt{\mu_1\varepsilon_1}\,\omega. \tag{4.45}$$

The wave vectors of the incident and reflected waves are now, respectively,

$$\vec{k}_i = k(\sin\theta_i, \cos\theta_i, 0) \tag{4.46a}$$

$$\vec{k}_r = k(\sin\theta_r, -\cos\theta_r, 0) \tag{4.46b}$$

with Poynting vectors

$$\vec{S}_I = \frac{1}{2}\frac{1}{\mu_1}\frac{1}{\omega}\mid E_I\mid^2\vec{k}_i \tag{4.47a}$$

$$\vec{S}_{\rm R} = \frac{1}{2}\frac{1}{\mu_1}\frac{1}{\omega} \mid E_{\rm R} \mid^2 \vec{k}_{\rm r} \tag{4.47b}$$

respectively. Equations (4.46) and (4.47) show the standard result of energy flow into and out of the interface of the two media. Since $\mu_1 < 0$ both incident and reflected wave Poynting vectors are anti-parallel to the wave vectors of their respective waves. This is a general property of propagation in a negative indexed medium.

For the present case of a positive indexed medium (i.e., ε_2, $\mu_2 > 0$) the transmitted wave in medium 2 is given by

$$\vec{E}_{\rm T} = E_{\rm T}(\cos\theta_{\rm t}\hat{i} + \sin\theta_{\rm t}\hat{j})e^{i[q(\sin\theta_{\rm t}x - \cos\theta_{\rm t}y) - \omega t]} \tag{4.48a}$$

$$\vec{B}_{\rm T} = E_{\rm T}\frac{q}{\omega}\hat{k}e^{i[q(\sin\theta_{\rm t}x - \cos\theta_{\rm t}y) - \omega t]} \tag{4.48b}$$

with the wave vector and Poynting vector of the transmitted wave, respectively, given by

$$\vec{q}_{\rm t} = q(\sin\theta_{\rm t}, -\cos\theta_{\rm t}, 0) \tag{4.49}$$

$$\vec{S}_{\rm T} = \frac{1}{2}\frac{1}{\mu_2}\frac{1}{\omega} \mid E_{\rm T} \mid^2 \vec{q}_{\rm t} \tag{4.50}$$

and where

$$q = \sqrt{\mu_2\varepsilon_2}\,\omega. \tag{4.51}$$

Since $\mu_2 > 0$ the wave vector and Poynting vectors are parallel, and the energy of the transmitted wave correctly flows away from the interface.

Again, the translational symmetry along the interface of the two media requires the x-components of the wave vectors of the incident, reflected, and transmitted waves to be equal. From this it follows that $\theta_{\rm i} = \theta_{\rm r}$ and $k\sin\theta_{\rm i} = q\sin\theta_{\rm t}$. Upon an application of the boundary conditions between the two media: (i) the continuity of the component of the electric field tangent to the interface requires

$$E_{\rm I} + E_{\rm R} = \alpha E_{\rm T} \tag{4.52a}$$

where $\alpha = \frac{\cos\theta_{\rm t}}{\cos\theta_{\rm i}}$ when medium 1 is negative indexed and medium 2 is positive indexed. (ii) At this same interface, the continuity of the components of the magnetic field tangent to the interface requires

$$-E_{\rm I} + E_{\rm R} = \beta E_{\rm T} \tag{4.52b}$$

where $\beta = \frac{\mu_1\sqrt{\mu_2\varepsilon_2}}{\mu_2\sqrt{\mu_1\varepsilon_1}}$. Solving these gives

$$E_{\rm T} = \frac{2}{\alpha - \beta}E_{\rm I} \tag{4.53a}$$

$$E_{\rm R} = \frac{\alpha + \beta}{\alpha - \beta}E_{\rm I}. \tag{4.53b}$$

4.3.3 Semiconductor analogy

The refraction of light at the interfaces between the positive and negative indexed media can be described in analogy with electrical current moving through an n–p or p–n semiconductor junction [9].

For light incident on the interface from the positive indexed medium an energy pulse is sent towards the interface. As the pulse travels to the interface it interacts with an energy hole coming towards the surface from the negative indexed medium. The hole, propelled in the negative indexed medium towards the interface, exists in response to the field from the incident pulse at the interface. At the interface the incident energy pulse and the hole destroy one another in the process of creating an energy pulse that is reflected from the surface into the positive indexed medium. The net result in the final state of the system is energy moving away from the surface in both the positive and negative indexed media.

For light incident on the interface from the negative indexed medium an energy hole travels away from the interface, creating an energy flow towards the interface. The field at the interface arising from the hole causes an energy pulse to be created at the interface. This pulse moves away from the interface and into the positive indexed medium. In addition, the fields at the interface create a hole traveling towards the interface through the negative indexed medium. The hole traveling towards the interface describes the energy reflected from the interface. The net result in the final state of the system is energy moving away from the interface and into both the positive and negative indexed media.

It is seen for both types of interfaces the energy transport properties are governed separately by the nature of the energy pulses in the positive medium and the nature of the energy holes in the negative indexed medium. The energy currents are parallel to the wave vector in the positive indexed medium and anti-parallel to the wave vector in the negative indexed medium. The energy pulses and holes in part destroy one another at the interface to give rise to various reflected and transmitted flows of energy.

4.3.4 The perfect lens

In traditional optics the refraction of light at the interface between different media is commonly used to make a focusing lens [8, 10]. The focusing lens directs light incident on it from an object in such a manner so as to create an image of the object. The resulting flow of light then appears to come from the created image rather than the original object. There are many problems in designing a lens that gives a good image, i.e., a nicely focused representation of the object. Some of these problems include: the curve of the surface, the dispersive properties of the media, and the presence of imperfections in the design and the materials [11].

One fundamental problem in the optics of positive indexed materials is that the refraction of a ray of light incident on a planar interface from the first or second quadrant can only be refracted into the third or fourth quadrants, respectively [3, 11]. A consequence of this is that to form a focusing lens from a positive index material requires at least one concave or convex curved surface. The curvature allows for the additional bending of the light ray which cannot be accomplished by the dielectric mismatch at a series of planar surfaces. As a result of the curvature of the lens surfaces, the two surfaces intersect one another in a circle of radius R which is the aperture of the lens. The finite size of the aperture of the lens is a fundamental limitation on the resolution of the image formed by the lens. Specifically, only wavelengths less that the diameter of the lens can leave the object, pass through the lens, and arrive at the focus of the lens. A precise criterion for the focusing power of a lens is that in forming an image of an object, only wavelengths of light, λ, satisfying

$$\lambda < R \tag{4.54}$$

are resolved in the focused image of the object. Such a lens is basically imperfect because of the limitation of its focusing ability arising from its finite R. To give a perfect image of the object, the image formed by the lens should contain all Fourier components of light rather than a restricted subset.

The new optics of negative indexed materials does not suffer from the necessity of making lenses of finite apertures. This follows from our earlier discussions of the refraction of light from a positive indexed medium to a negative indexed medium. In the optics of negative indexed materials the refraction of a ray of light incident on a planar interface from the first or second quadrant of a positive indexed medium is refracted into the fourth or third quadrants of the second negative indexed medium, respectively. This means that the additional beam bending arising from a curved surface between the two media is not necessary. In addition, from our earlier discussions of the refraction of light from a negative indexed medium to a positive indexed medium: the refraction of a ray of light incident on a planar interface from the first or second quadrant of a negative indexed medium is refracted into the fourth or third quadrants of the second positive indexed medium, respectively. This again precludes the need for additional bending from a curved surface.

In principle a slab of negative indexed medium surrounded by a positive indexed medium can form a lens with perfect resolution. The system can be used to focus an image–object pair in the positive indexed medium on the opposite sides of the slab. Since the slab is of infinite extent with an infinite aperture (i.e., $R \to \infty$) the lens is able to focus all propagating wavelength components of the object in its image. In addition the permittivity and permeability can be adjusted so that even the evanescent components of light from the object are reassembled at the image, giving a complete characterization in the image of the object [10]. In principle an image with perfect resolution can be formed by imaging with an infinite slab of negative index medium. The lens in this sense is a perfect lens. It is important to note, however, that due to the magnetic resonance origins of engineered negative indexed materials, the perfect lens model has some additional complications in its experimental realization. Resonances with negative permeabilities only exist over a band of frequencies. Also, resonances are associated with dielectric losses which become greater near the frequency peak of the resonance. Consequently, the idea of a perfect lens has only been very narrowly realized experimentally.

In addition, negative indexed media can allow for the development of an electromagnetic cloaking. Using a gradual spatial variation of the refractive index in an engineered medium, light can be steered around an object hidden in the medium and sent off parallel to its original direction before it encountered the cloaking device [8]. To do this requires the use of both positive and negative indexed media in order to accomplish all of the bending of light required of the cloaking device. Due to the restrictions from magnetic resonance effects, this has only been done for the case of a single frequency of incident light. Using the idea of a medium with a spatial variation of positive and negative index, suggestions for the mimicking of certain optical effects in general relativity have also been proposed [12–14].

4.3.5 Radiation in negative indexed media

Another example of new effects associated with negative indexed media involves the properties of radiation fields generated within them. In this subsection, the radiation fields of an electric dipole antenna located in a negative indexed medium and of a point change moving within a negative index medium are treated [5, 15]. An important determiner of the properties of the radiation fields comes from the fact that the wave vector and Poynting vector of the radiation in a negative indexed medium are anti-parallel.

In the Lorentz gauge (i.e., $\nabla \bullet \vec{A} = -\frac{1}{c_{\mathrm{m}}^2}\frac{\partial V}{\partial t}$) the radiation equations for electromagnetic waves generated from a source are [15]

$$\nabla^2 V - \frac{1}{c_{\mathrm{m}}^2}\frac{\partial^2 V}{\partial t^2} = -\frac{1}{\varepsilon}\rho \tag{4.55a}$$

$$\nabla^2 \vec{A} - \frac{1}{c_{\mathrm{m}}^2}\frac{\partial^2 \vec{A}}{\partial t^2} = -\mu\vec{J} \tag{4.55b}$$

where $(\vec{A}(\vec{r}, t), V(\vec{r}, t))$ are the vector and scalar potentials and $(\vec{J}(\vec{r}, t), \rho(\vec{r}, t))$ are the current and charge densities of the source and c_{m} is the speed of light in the medium. The solutions of (4.55) are

$$V_{\mp}(\vec{r}, t) = \frac{1}{4\pi\varepsilon}\int \frac{\rho(\vec{r}\,', t_{\mp})}{|\vec{r} - \vec{r}\,'|}\mathrm{d}^3r' \tag{4.56a}$$

$$\vec{A}_{\mp}(\vec{r}, t) = \frac{\mu}{4\pi}\int \frac{\vec{J}(\vec{r}\,', t_{\mp})}{|\vec{r} - \vec{r}\,'|}\mathrm{d}^3r' \tag{4.56b}$$

where $t_{\mp} = t \mp \frac{|\vec{r} - \vec{r}\,'|}{c_{\mathrm{m}}}$. In (4.56) the upper (lower) signs are known as the retarded (advanced) potentials of the fields generated by the source terms. For positive indexed media the radiation fields describing the motion of radiation away from the sources are obtained from the retarded potentials. Later, we shall see that the advanced potentials are those of interest for radiation generated in a negative indexed medium.

The electric field and the magnetic induction are then obtained from the vector and scalar potentials as

$$\vec{E} = -\nabla V - \frac{\partial \vec{A}}{\partial t} \tag{4.57a}$$

and

$$\vec{B} = \nabla \times \vec{A} \tag{4.57b}$$

respectively. Note that in (4.56) as $c_m \to \infty$ the potentials, appropriately, reduce to the scalar and vector potentials of static electric and magnetic fields.

As an example of the formulation consider an electric point dipole source located at the origin of coordinates and having the form of a harmonic time-dependent electric dipole given by $\vec{p}(t) = p_0\cos(\omega t)\hat{k}$. In the far field limit in which the point of observation is a great distance (i.e., $kr \gg 1$) from the source, the electric field and magnetic induction generated by the time-dependent dipole are

$$\vec{E} = -\frac{\mu p_0 \omega^2}{4\pi}\frac{\sin\theta}{r}\cos\omega\left(t \mp \frac{r}{c_m}\right)\hat{\theta} \tag{4.58a}$$

$$\vec{B} = \mp\frac{\mu p_0 \omega^2}{4\pi c_m}\frac{\sin\theta}{r}\cos\omega\left(t \mp \frac{r}{c_m}\right)\hat{\varphi} \tag{4.58b}$$

where (r,θ,φ) are the standard polar coordinates centered at the origin of coordinates. From (4.58) the Poynting vector of the radiation fields is

$$\langle\vec{S}\rangle = \pm\frac{\mu p_0^2 \omega^4}{32\pi^2 c_m}\frac{\sin^2\theta}{r^2}\hat{r} \tag{4.59}$$

giving an average power

$$\langle P\rangle = \pm\frac{\mu}{12\pi}\frac{p_0^2\omega^4}{c_m}. \tag{4.60}$$

For a positive indexed medium the permeability is positive and the upper sign from the retarded solution gives an energy flow away from the dipole source. For a negative indexed medium the permeability is negative and the lower sign from the advanced solution gives an energy flow away from the dipole source.

An additional important case is that of the determination of the radiation from a point charge accelerating in a dielectric medium. To handle this system, first consider the radiation characteristics of a general time-dependent localized charge distribution, given by (4.56). The results for a general charge and current density are then immediately specialized to the problem of a single accelerating point charge. In the case that $r' \ll r$, for a charge distribution located about the origin of coordinates, the non-relativistic limit of (4.56) becomes

$$V_{\mp}(\vec{r}, t) = \frac{1}{4\pi\varepsilon}\left[\frac{Q}{r} + \frac{\hat{r} \bullet \vec{p}(t_{\mp})}{r^2} + \frac{\hat{r} \bullet \dot{\vec{p}}(t_{\mp})}{c_m r}\right] \tag{4.61a}$$

$$\vec{A}_{\mp}(\vec{r}, t) = \frac{\mu}{4\pi}\frac{\vec{\dot{p}}(t_{\mp})}{r} \quad . \tag{4.61b}$$

Here Q is the net charge of the distribution and $\vec{p}(t)$ is the electric dipole moment of the charge distribution about the origin of coordinates.

For the accelerating point charge it is only needed to determine the electric dipole moment of the charge relative to the origin of coordinates. It then follows that the radiation fields of the charge distribution are

$$\vec{E}(r, \theta, \varphi, t_{\mp}) = \frac{\mu}{4\pi}\vec{\ddot{p}}(t_{\mp})\frac{\sin\theta}{r}\hat{\theta} \tag{4.62a}$$

$$\vec{B}(r, \theta, \varphi, t_{\mp}) = \pm\frac{\mu}{4\pi}\frac{\vec{\ddot{p}}(t_{\mp})}{c_{\mathrm{m}}}\frac{\sin\theta}{r}\hat{\varphi} \tag{4.62b}$$

where (r, θ, φ) are standard polar coordinates chosen so that the charge is accelerating along the z-axis. The Poynting vector of these radiation fields is

$$\vec{S} = \pm\frac{\mu[\ddot{p}(t)]^2}{16\pi^2 c_{\mathrm{m}}}\frac{\sin^2\theta}{r^2}\hat{r} \tag{4.63}$$

giving a net radiated power

$$P = \pm\frac{\mu}{6\pi}\frac{q^2 a^2(t)}{c_{\mathrm{m}}} \tag{4.64}$$

where $a(t)$ is the acceleration of the charge. For a positive indexed medium the permeability and permittivity are positive. In this case, the upper sign from the retarded solution gives an energy flow away from the dipole source. For a negative indexed medium the permeability and permittivity are negative and the lower sign from the advanced solution gives an energy flow away from the dipole source. Again, in the positive indexed medium the Poynting vector and wave vector are parallel while they are anti-parallel in the negative indexed medium.

References

[1] Cai W and Shalaev V 2001 *Optical Metamaterials: Fundamental and Applications* (New York: Springer)

[2] Engheta N and Ziolkowski R W (ed) 2006 *Metamaterials: Physics and Engineering Explorations* (New York: Wiley-IEEE)

[3] Ramakrishna S A 2005 Physics of negative index materials *Rep. Prog. Phys.* **68** 449

[4] Veselago V G 1968 The electrodynamics of substances with simultaneously negative values of ε and μ *Sov. Phys.– Usp.* **10** 509

[5] Agranovich V M 2006 Hybrid organic–inorganic nanostructures and light–matter interaction *Problems of Condensed Matter Physics* ed A L Ivanov and S G Tikhodeev (Oxford: Clarendon) ch 2

[6] Eleftheriou M, Lazarides N and Tsironis G P 2008 Magnetoinductive breathers in metamaterials *Phys. Rev.* E **77** 036608

[7] Eleftheriades G V 2009 EM transmission-line metamaterials *Mater. Today* **12** 30
[8] Kourakis I, Lazarides N and Tsironis G P 2007 Self-focusing and envelope pulse generation in nonlinear magnetic metamaterials *Phys. Rev.* E **75** 067601
[9] Marder M P 2010 *Condensed Matter Physics* 2nd edn (Hoboken, NJ: Wiley)
[10] Pendry J B 2000 Negative refraction makes a perfect lens *Phys. Rev. Lett.* **85** 3966
[11] Smith W J 2007 *Modern Optical Engineering* 4th edn (New York: McGraw-Hill)
[12] Leonhardt U and Philibin T G 2006 General relativity in electrical engineering *New J. Phys.* **8** 247
[13] Leonhardt U and Philibin T G 2009 Transformation optics and the geometry of light *Prog. Opt.* **52** 69
[14] Van B J 1984 *Relativity and Engineering* (Berlin: Springer)
[15] Jackson J D 1999 *Classical Electrodynamics* 3rd edn (New York: Wiley)

Chapter 5

Nonlinearity in photonic crystal and metamaterial waveguides

In this chapter the formulation of the problem of Kerr nonlinearity in photonic crystals and metamaterials is discussed. The Kerr nonlinearity causes the dielectric properties of the system to depend on the intensity of the electric fields applied to it [1–4]. This allows for the modification of the properties of the electromagnetic excitations found within the system. The formulation for both photonic crystals and metamaterials will express the dynamical equations as sets of difference equations [5–15]. In addition, in the continuum limit of these difference equations, a dynamics involving partial differential equations will be developed [5–11, 16].

The effects of Kerr nonlinearity have been studied in a variety of photonic crystals and metamaterials. These include the treatment of bulk three-dimensional systems composed of Kerr media, waveguides of Kerr media introduced within systems otherwise composed of linear dielectric media, and site and clusters of site Kerr nonlinear impurities within bulk linear systems. To restrict the considerations, a focus will be on waveguide structures. In general, these are easier to treat than bulk systems, while illustrating some of the basic dynamics of nonlinear many-body structures in photonic crystals and metamaterials. The dynamical properties of nonlinear structures can be determined by both computer simulation and analytic techniques. Some of the results from computer simulation techniques will be discussed as well as the results of analytical methods, but the treatment will first focus on analytical methods [5–11, 13–16].

5.1 SRR metamaterials

Let us begin by writing the equations that describe the dynamics of a chain of linear medium split ring resonators (SRRs) [5–11, 13, 14]. This can be thought of as being a simple waveguide or transmission line. The SRRs in the chain are composed of linear dielectric media and are arrayed on a simple one-dimensional Bravais lattice. The SRRs along the chain are coupled by mutual inductance between different SRRs with the separation between the SRRs great enough that only nearest neighbor couplings are considered. The equation of motion of the system is given by [5–8]

$$L\frac{d^2Q_n}{dt^2} + \frac{Q_n}{C_l} + M\left[\frac{d^2Q_{n+1}}{dt^2} + \frac{d^2Q_{n-1}}{dt^2}\right] = 0 \tag{5.1}$$

where Q_n is the charge on the capacitor of the nth SRR, C_l is the capacitance of each SRR, L is the SRR self-inductance, and M is the weak nearest neighbor mutual inductance between the SRRs. For $M = 0$ (5.1) reduces to the standard equation of motion of a simple LC circuit, encountered in elementary physics. The last term on the left-hand side represents the inductive couplings between neighboring SRR circuits arising from Faraday's law interactions.

Solutions of (5.1) are obtained in terms of plane waves of the form $Q_n = Qe^{i(kn-\omega_l t)}$ and a substitution into (5.1) gives the dispersion relation of the plane wave modes as

$$-\omega_l^2[L + 2M\cos k] + \frac{1}{C_l} = 0. \tag{5.2}$$

The general solutions for the dynamics of the chain are given as linear combinations of the plane wave modes.

Next consider the modifications of (5.1) arising from the introduction of a Kerr nonlinear dielectric into the SRR capacitance. The Kerr nonlinear dielectric constant is taken to be of the form [5]

$$\varepsilon(|E|^2) = \varepsilon_0\left(\varepsilon_l + \alpha\frac{|E|^2}{E_c^2}\right) \tag{5.3}$$

where $\alpha = \pm 1$ is the sign of the Kerr interaction, and E_c sets the strength of the Kerr interaction. In the limit that $\alpha = 0$ (5.3) reverts to the linear dielectric constant of the system treated in (5.1). For the nonlinear medium system, to leading order in the small nonlinear term of the Kerr dielectric (5.1) becomes [5]

$$L\frac{d^2Q_n}{dt^2} + \frac{Q_n}{C_l} + M\left[\frac{d^2Q_{n+1}}{dt^2} + \frac{d^2Q_{n-1}}{dt^2}\right] - \frac{\alpha}{\varepsilon_l U_c^2}\left(\frac{Q_n}{C_l}\right)^3 = 0 \tag{5.4}$$

where the last term on the left-hand side describes the weak nonlinear Kerr effects. In terms of the E_c defined in (5.3), $U_c = E_c d$ where d is the capacitive gap of the SRR, and the C_l in (5.4) is the capacitance of an SRR in the $E_c \to \infty$ limit. The Q_n^3 correction comes from the change in capacitor energy as the electric field in the capacitor changes, and the details of the determination of this term are found in the literature [5]. It is important to note that in applying (5.3) to obtain (5.4), the absolute value $|E|$ in (5.3) has been replaced by the field E. Equation (5.3) in its present form expresses an application of what is called the rotating wave approximation (RWA). This is an approximation which effectively decouples modes of different frequencies and is standard to the formulation of the Kerr dielectric. Some explanation of this is now presented.

An important point to note is that the Q_n^3 term in (5.4) introduces a mixing of the plane wave solutions of the linear limit of the system. Specifically, if Q_n contains frequency components $e^{i\omega t}$ and $e^{-i\omega t}$, then Q_n^3 contains $e^{\pm i\omega t}$, $e^{i3\omega t}$, and $e^{-i3\omega t}$ terms in addition to those of the original fundamental frequency. The solution of the system, then, involves a mixing of many frequency components. We shall only be interested in the weak nonlinear case and ignore the mixing of frequency components. This is known as the RWA and involves replacing Q_n^3 by a term proportional to $|Q_n|^2 Q_n$. This will be further discussed later.

In terms of dimensionless units (5.4) becomes [5]

$$\ddot{q}_n + \lambda\left(\ddot{q}_{n+1} + \ddot{q}_{n-1}\right) + q_n - \gamma q_n^3 = 0 \tag{5.5}$$

where $\lambda = \frac{M}{L}$, $\gamma = \frac{\alpha}{\varepsilon_l}$, $q_n = \frac{Q_n}{C_l U_c}$, with $\ddot{q}_n = \frac{d^2 q_n}{d\tau^2}$, and $\tau = \frac{t}{\sqrt{LC_l}}$. For further details on this normalization see [5]. It shall be seen later that the addition of the nonlinear last term on the left-hand side of (5.5) does more than just renormalize the plane wave solutions of the linear chain. It also leads to additional soliton-like modes in the system. In addition, a linear combination of solutions of (5.5) is no longer a new solution of the system. This is a fundamental difference between the solutions of linear and nonlinear systems. The solutions of (5.5) will be discussed in a later section.

A limit of (5.5) which is often quite helpful in understanding the properties of the solutions of (5.5) is the continuum limit. In this case, the wave functions of the solutions of (5.5) change slowly from site to neighboring site along the chain. In the continuum limit for which the solutions of (5.5) are slowly changing along the chain (5.5) can be rewritten as a differential equation. For the continuum limit we then take

$$\ddot{q}_{n\pm 1} \approx \left[\ddot{q} \pm a\frac{\partial \ddot{q}}{\partial x} + \frac{1}{2}a^2\frac{\partial^2 \ddot{q}}{\partial x^2}\right]_{x=na}. \tag{5.6a}$$

Here $q(x)$ on the right-hand side of the equation is now a continuous variable of position along the chain, the chain is taken to lie along the x-axis, and a is the lattice constant. At the lattice sites, $q(x)$ equals the discrete lattice values q_n, and it gives a smooth interpolation between neighboring lattice sites. Using (5.6a) and assuming a small amplitude wave, (5.5) becomes [5]

$$\frac{\partial^2 q}{\partial \tau^2} - cb\frac{\partial^2 q}{\partial x^2} + b[q - \gamma q^3] = 0 \tag{5.6b}$$

where $c = \lambda/(1 + 2\lambda)$ and $b = 1/(1 + 2\lambda)$.

Equation (5.6b) is a nonlinear Klein–Gordon equation, and its solutions will be discussed later. Here it is noted, however, that most discussions of optical nonlinear systems are made in terms of a formulation involving the nonlinear Schrödinger equation rather than the nonlinear Klein–Gordon equation. The nonlinear Schrödinger equation [5–11, 14, 16] is of the general form

$$\mathrm{i}\frac{\partial q}{\partial \tau} + P\frac{\partial^2 q}{\partial \zeta^2} + Q\,|q|^2\,q = 0 \tag{5.7}$$

where (ζ, τ) are space–time coordinates and for real coefficients P and Q. Equation (5.7) is basically the linear Schrödinger equation with a cubic nonlinearity. It can be shown that the nonlinear Schrödinger equation is closely related to the more general nonlinear Klein–Gordon equation. The nonlinear Schrödinger equation is obtained from the nonlinear Klein–Gordon equation by applying multiple scale perturbation theory and the RWA to the Klein–Gordon equation. Later the solutions of the nonlinear Schrödinger equation will also be discussed.

5.2 Photonic crystal waveguides

Consider a bulk two-dimensional photonic crystal with electromagnetic waves propagating in the plane of the periodicity. The fields of the waves satisfy Helmholtz equations

$$[-\nabla^2 - \mu\varepsilon(\vec{r})\omega^2]\vec{E}(\vec{r},\,\omega) = 0 \tag{5.8a}$$

$$[-\nabla^2 - \mu\varepsilon(\vec{r})\omega^2]\vec{B}(\vec{r},\,\omega) = 0 \tag{5.8b}$$

where $\varepsilon(\vec{r})$ is the periodic permittivity and $\vec{r}$ is in the plane of the periodicity. A waveguide or impurity can be introduced into the system by adding to $\varepsilon(\vec{r})$ a position-dependent change in the dielectric constant, $\delta\varepsilon(\vec{r})$. The effects of $\delta\varepsilon(\vec{r})$ are then determined by employing standard Green's function techniques, similar to those used in scattering theory. Specifically, for electric fields polarized perpendicular to the plane of periodicity, the fields in the presence of the $\delta\varepsilon(\vec{r})$ satisfy [12]

$$E(\vec{r},\,\omega) = \left(\frac{\omega}{c}\right)^2 \int \mathrm{d}^2 r'\, G(\vec{r},\,\vec{r}'|\,\omega)\delta\varepsilon(\vec{r}')E(\vec{r}',\,\omega) \tag{5.9}$$

where $G(\vec{r},\vec{r}'|\omega)$ is the appropriately defined Green's function of the Helmholtz operators on the left-hand side of (5.8). In the following the focus will be on (5.9) and its solutions, but the fields of the other polarization are obtained from an equation of a similar form. Specifically, results for the polarization with the electric field in the plane of propagation are matrix equations that can be found in the literature.

Equation (5.9) is similar in form to field equations for atomic or nuclear scattering or to field equations describing electronic impurities in metals and semiconductors [17]. The equations in these problems are generally very difficult to treat as their scattering potentials must be treated approximately. For the photonic crystal problem this is not the case, and, in fact, $\delta\varepsilon(\vec{r})$ is known to be a constant in space or a periodically repeated constant in space. In this case (5.9) can often be viewed as a type of eigenvalue problem for the eigenvector $E(\vec{r},\omega)$ and the constant eigenvalue of $\varepsilon(r)$ obtained for a fixed ω^2. This offers a simplified solution.

As an example [12], consider the case in which there is a single impurity site in a bulk two-dimensional photonic crystal. Let $\delta\varepsilon(\vec{r}) = \delta\varepsilon_{00}$ over some connected region, S, in the plane of the periodicity and $\delta\varepsilon(\vec{r}) = 0$ otherwise, then from (5.9) we find for $\vec{r}\,' \in S$

$$E(\vec{r}, \omega) = \left(\frac{\omega}{c}\right)^2 \delta\varepsilon_{00} \int d^2r' G(\vec{r}, \vec{r}\,'| \, \omega) E(\vec{r}\,', \omega). \tag{5.10}$$

If $E(\vec{r}, \omega)$ is slowly varying over S, then $\left(\frac{\omega}{c}\right)^2 \int d^2r' G(\vec{r}, \vec{r}\,'| \, \omega) E(\vec{r}\,', \omega) \approx \alpha(\vec{r}, \vec{r}\,') E(\vec{r}\,', \omega)$ where $\vec{r}\,' \in S$ and $\alpha(\vec{r}, \vec{r}\,') = \left(\frac{\omega}{c}\right)^2 \int_S d^2r' G(\vec{r}, \vec{r}\,'| \, \omega)$. In the case that $\vec{r} = \vec{r}\,'$

$$[1 - g\alpha] E_0 = 0, \tag{5.11}$$

where $\alpha = \alpha(\vec{r}, \vec{r})/S$, $g = S\delta\varepsilon_{00}$, S is the area of the impurity in the plane of the periodicity, and $E_0 = E(\vec{r}, \omega)$. For frequencies chosen in the stop bands of the photonic crystal, this provides an eigenvalue solution for the bound state modes of the single impurity problem. In particular, treating g as the unknown, (5.11) is solved for g in terms of the other known variables of the equations evaluated at the chosen stop band frequency of the impurity level. A similar problem is encountered in the study of n- and p-bound state impurities in electronic semiconductors [17].

Similarly, a difference equation formulation is obtained from (5.9) for a waveguide composed by cylinder replacement as a linear chain of periodically repeated impurity sites along an axis of the photonic crystal. Specifically, considering a photonic crystal of dielectric cylinders defined on a two-dimensional square lattice with nearest neighbor sites oriented along the x- and y- axes, for a waveguide channel along the x-axis [12] the difference equation arising from (5.9) takes the general form

$$E_n = g\,[\alpha E_n + \beta\,(E_{n-1} + E_{n+1})]. \tag{5.12}$$

Here the system is limited to on site and nearest neighbor couplings α and β, respectively. (Note that $\beta = \alpha(\vec{r}, \vec{r}\,')/S$ where $\vec{r}$ and $\vec{r}\,'$ are in neighboring replacement sites in the waveguide channel.) The fields are labeled with a site index, consecutively numbering the sites on the chain. For studying waveguide modes the frequency in (5.12) is set in a stop band of the bulk photonic crystal so that the neighbor couplings decay exponentially with the separation of the waveguide sites. For far enough nearest neighbor separations nearest neighbor couplings are all that are needed, and the discussions later will be restrict to this case. In the presence of a Kerr nonlinear permittivity $\delta\varepsilon(\vec{r}) = \delta\varepsilon_{00}(1 + \lambda\,|E|^2)$ where E is the applied electric field at the medium and λ is its Kerr parameter, (5.12) becomes [14, 15, 18]

$$E_n = g\left[\alpha\left(1 + \lambda\,|E_n|^2\right)E_n + \beta\left(1 + \lambda\,|E_{n+1}|^2\right)E_{n+1} + \beta\left(1 + \lambda\,|E_{n-1}|^2\right)E_{n-1}\right]. \tag{5.13}$$

The modes of (5.12) and (5.13) will be discussed later.

The photonic crystal waveguide equation for modes slowly varying in space can also be handled within the continuum limit. In the continuum limit the position-dependent electric field is expanded as

$$E_{n\pm1} \approx \left[E \pm a\frac{\partial E}{\partial x} + \frac{1}{2}a^2\frac{\partial^2 E}{\partial x^2}\right]_{x=na} \tag{5.14}$$

where a is the lattice constant of the waveguide channel. For a slowly varying field envelope and a small Kerr parameter (5.13) reduces to a nonlinear differential equation of the form

$$A\frac{\partial^2 E}{\partial x^2} + [B + C\lambda\,|E|^2]\,E = 0 \tag{5.15}$$

where $A = g\beta a^2$, $B = g\,(\alpha + 2\beta) - 1$, and $C = g(\alpha + 2\beta)$. The solutions of this nonlinear differential equation will be discussed later.

In the following chapter the nonlinear differential equations and their soliton solutions will be discussed first. These types of equations have traditionally been the focus of much attention. After this the solutions of the difference equation formulations will be approached. The results of the two methods will then be compared.

References

[1] Boyd R W 2003 *Nonlinear Optics* 2nd edn (Amsterdam: Academic)

[2] Mills D L 1998 *Nonlinear Optics* (Berlin: Springer)

[3] Banerjee P P 2004 *Nonlinear Optics* (New York: Dekker)

[4] Kivshar Y S and Agrawal G P 2003 *Optical Solitons* (Amsterdam: Academic)

[5] Giri P, Choudhary K, Sen Gupta A, Bandyopadhyay A K and McGurn A R 2011 Klein–Gordon equation approach to nonlinear split-ring resonator based metamaterials: one-dimensional systems *Phys. Rev.* B **84** 155429

[6] Eleftheriou M, Lazarides N and Tsironis G P 2008 Magnetoinductive breathers in metamaterials *Phys. Rev.* E **77** 036608

[7] Eleftheriades G V 2009 EM transmission-line metamaterials *Mater. Today* **12** 30

[8] Kourakis I, Lazarides N and Tsironis G P 2007 Self-focusing and envelope pulse generation in nonlinear magnetic metamaterials *Phys. Rev.* E **75** 067601

[9] Lapine M, Gorkunov M and Ringhofer K H 2003 Nonlinearity of a metamaterial arising from diode inserts into resonant conductive elements *Phys. Rev.* E **67** 065601

[10] Shadrivov I V, Morrison S K and Kivshar Y 2006 Tunable split-ring resonators for nonlinear negative-index metamaterials *Opt. Express* **14** 9344

[11] Tamayama Y, Nakanishi T and Kitano M 2013 Observation of modulation instability in a nonlinear magnetoinductive waveguide *Phys. Rev.* B **87** 195123

[12] McGurn A R 1996 Green's-function theory for row and periodic defect arrays in photonic band structures *Phys. Rev.* B **53** 7059

[13] Sievers A J and Page J B 1995 *Dynamical Properties of Solids* ed G K Horton and A A Maradudin (Amsterdam: North-Holland)

[14] McGurn A R 2013 Barrier transmission map of one-dimensional nonlinear split-ring-resonator-based metamaterials: bright, dark, and gray soliton resonances *Phys. Rev.* B **88** 155110

[15] McGurn A R 2008 Transmission through nonlinear barriers *Phys. Rev.* **B 77** 115105
[16] Dauxois T and Peyrard M 2006 *Physics of Solitons* (Cambridge: Cambridge University Press)
[17] Galin J S 2000 *Impurity Scattering in Metallic Alloys* (New York: Springer)
[18] McGurn A R 1999 Intrinsic localized modes in nonlinear photonic crystal waveguides *Phys. Lett.* A **251** 322

IOP Publishing

Introduction to Nonlinear Optics of Photonic Crystals and Metamaterials (Second Edition)

Arthur R McGurn

Chapter 6

Nonlinear differential equations in physics and their soliton solutions

The nonlinear differential equations of nano-systems fall into the forms of the nonlinear Klein–Gordon and nonlinear Schrödinger equations [1]. In the following these equations are discussed and their solutions are examined. It is shown that the solutions of both equations are classified as plane wave and soliton solutions, and the distinct propagation characteristics between these two types of solutions are contrasted. This will be followed by a study of some results from their applications in nano-photonic systems.

The nonlinear Klein–Gordon and nonlinear Schrödinger equations are obtained from the linear Helmholtz and the linear Schrödinger equations through the addition to these linear forms of terms dependent on nonlinear powers of the wave function [2]. It is assumed that the medium of the bulk material in which the waveguide is embedded is composed of linear optical media so that the added nonlinear terms come from the nonlinearity of the optical media within the waveguide. The nonlinearity of the media in the waveguide is treated in the rotating wave approximation (RWA) and described by a Kerr dielectric constant which depends on the intensity of the fields applied to it. Nonlinearity enters the waveguide equations as terms involving second powers of the absolute values of the wave function or field amplitudes.

For a waveguide composed of linear media in the one-dimensional geometry of a straight waveguide situated along the x-axis, the field amplitudes of the guided waves along the chain can often be expressed in terms of the solutions of the linear Helmholtz or the linear Schrödinger equations. As we shall see later, which of these two equations are used depends on the level of the approximation considered in going from the discrete model of a waveguide system to its continuum limit. The Helmholtz equation describing the field amplitude is of the form

$$\frac{\partial^2\psi}{\partial x^2} - \frac{1}{v^2}\frac{\partial^2\psi}{\partial t^2} = 0 \tag{6.1}$$

where v is the phase velocity of the field and $\Psi(x,t)$ is the appropriate field amplitude. Likewise, the linear Schrödinger equation is of the general form

$$-\frac{\hbar^2}{2m}\frac{\partial^2\psi}{\partial x^2} - \mathrm{i}\hbar\frac{\partial\psi}{\partial t} = 0 \tag{6.2}$$

where $\frac{\hbar^2}{2m}$ and $\mathrm{i}\hbar$ are constants which have been formatted to explicitly give the appearance, familiar from quantum mechanics, of the Schrödinger equation. Both equations have solutions of the general form

$$\psi(x, t) = f(kx - \omega t) \tag{6.3}$$

where $f(y)$ is a twice differentiable function, and the parameters k and ω are determined by substitution in (6.1) or (6.2).

Often $f(kx-\omega t)$ is chosen to be one of the set of orthogonal plane waves $\{e^{\mathrm{i}(kx-\omega t)}\}$. The set and its orthogonality properties are determined from the boundary conditions of the waveguide problem so that the resulting set forms a complete basis. A general solution of (6.1) and (6.2) is then expressed as a linear combination in the basis. For example, substituting $e^{\mathrm{i}(kx-\omega t)}$ into (6.1) gives

$$\left[k^2 - \frac{1}{v^2}\omega^2\right]e^{\mathrm{i}(kx-\omega t)} = 0 \tag{6.4}$$

for the plane wave solution. The dispersion relation is then $\frac{\omega}{k} = \pm v$, and the appropriate wave vector set, $\{k\}$, is determined by the boundary conditions. A similar substitution of the plane wave set into (6.2) gives a dispersion relation and set of wave vectors for an orthogonal basis set of that equation. The linearity of the wave equations allows for a general solution of (6.1) and (6.2) to be expressed as a linear combination of their orthonormal plane wave bases.

In contrast to the simple properties of the solutions of the linear systems discussed above, fundamental complications are found to arise in the behaviors of (6.1) and (6.2) with the addition to them of nonlinear terms. A fundamental complication found with the addition of nonlinearity is that a linear combination of solutions is no longer another solution of the system. This hinders the systematic development of the spectrum of solutions. It also enables the nonlinear systems to exhibit new soliton-like solutions which are absent in the solution sets of their linear limits, i.e., the set of solutions of nonlinear systems is increased over that of their linear system limiting forms. Upon investigating the effects of nonlinearity using perturbation methods a set of plane wave solutions of the nonlinear equations can be developed from the corresponding plane waves of the linear limit of the systems. Soliton solutions of the nonlinear equations, however, do not arise in the perturbation treatment of the nonlinearity. Solitons are fundamentally new solutions of the nonlinear systems which do not originate from excitations found in the linear limit of the system.

The nonlinear Klein–Gordon and Schrödinger equations are obtained from (6.1) and (6.2) by the addition of terms to the potential of the equation which depend on powers of the intensity of the wave function, i.e., potential terms involving $|\psi(x, t)|$. The dependence on the intensity limits the solution set to include waves of a single frequency as it does not include terms which couple to or generate higher or lower harmonics to the fundamental of the system. Specifically, it decouples the harmonics from each other. In one dimension the resulting nonlinear Klein–Gordon equation, obtained by adding terms linear and nonlinear in $\Psi(x, t)$ to the Helmholtz equation, is of the form

$$\frac{\partial^2\psi}{\partial x^2} - \frac{1}{v^2}\frac{\partial^2\psi}{\partial t^2} + A\psi + B \mid \psi \mid^2 \psi = 0 \tag{6.5}$$

where the added nonlinear term is said to be in the RWA so that harmonics are not mixed by the equation. The nonlinear Schrödinger equation is similarly obtained from the Schrödinger wave equation as

$$-\frac{\hbar^2}{2m}\frac{\partial^2\psi}{\partial x^2} - i\hbar\frac{\partial\psi}{\partial t} + A \mid \psi \mid^2 \psi = 0 \tag{6.6}$$

through the additions of a cubic term which is again said to be in the RWA. These two wave equations both have solutions of the form

$$\psi(x, t) = f(kx - \omega t), \tag{6.7}$$

where now $f(y)$ has a restricted functional form determined by the particulars of its nonlinear equation and the parameters k and ω are determined by (6.5) or (6.6) under appropriate waveguide boundary conditions. Unlike the linear wave equation solutions, the exact form of the solution in (6.7) is determined by the specifics of the nonlinear equation it solves, but it is found to be in general classified as a plane-wave-like solution or a soliton or multiple-soliton-like solution. These solution types are now discussed.

The solutions of the nonlinear equations are no longer composed as linear combinations of a plane wave basis, but the equations do have plane wave solutions. The simple plane waves of the set $\{e^{i(kx-\omega t)}\}$ for a specified wave amplitude each, individually, represent solutions of (6.5) and (6.6). Unlike the plane wave solutions of the linear equations, these solutions are found to have dispersion relations which depend on the amplitude of the plane wave solutions. This dependence of the dispersion relation on the amplitude of the plane waves is the new feature of nonlinear equations which is responsible for the inability of forming new solutions as linear combinations of individual plane wave solutions. Substituting a plane wave form $\psi_0 e^{i(kx-\omega t)}$ into (6.5) we find

$$-k^2 + \frac{1}{v^2}\omega^2 + A + B \mid \psi_0 \mid^2 = 0 \tag{6.8}$$

which is solved for the dispersion relation of the frequency $\omega(k, \psi_0)$ as a function of the wave vector and amplitude of the plane wave. Likewise substitution of the plane wave form into (6.6) gives

$$\frac{\hbar^2 k^2}{2m} - \hbar\omega + A \mid \psi_0 \mid^2 = 0 \tag{6.9}$$

for the dispersion relation $\omega(k, \psi_0)$ as a function of the wave vector and amplitude of the plane wave.

In addition to the set of plane wave solutions, there are new types of solutions not found in systems described by linear equations. The new types of excitations only occur in nonlinear systems and include a variety of soliton and multi-soliton modes possessing distinctly different characteristics from those of the plane wave solutions [1]. The properties of these new excitations will now be the focus of the discussions.

To study the soliton-like solutions of (6.5) and (6.6), begin by looking for solutions to these equations written in the form of a plane wave modulated by a function of x and t. The wave function is then expressed as

$$\psi(x, t) = \mathrm{e}^{\mathrm{i}(kx-\omega t)} g(x, t). \tag{6.10}$$

Substituting (6.10) into (6.5) results in the equation

$$\frac{\partial^2 g}{\partial x^2} - \frac{1}{v^2}\frac{\partial^2 g}{\partial t^2} + 2\mathrm{i}\left[k\frac{\partial g}{\partial x} + \frac{\omega}{v^2}\frac{\partial g}{\partial t}\right] + \left[-k^2 + \frac{\omega^2}{v^2} + A + B \mid g \mid^2\right]g = 0 \tag{6.11}$$

giving the Klein–Gordon equation modulation $g(x,t)$ as a function of x and t. A similar substitution into (6.6) gives an equation for $g(x,t)$ of the Schrödinger equation having the form

$$-\frac{\hbar^2}{2m}\frac{\partial^2 g}{\partial x^2} + \mathrm{i}\left[-\frac{\hbar^2 k}{m}\frac{\partial g}{\partial x} - \hbar\frac{\partial g}{\partial t}\right] + \left[\frac{\hbar^2 k^2}{2m} - \hbar\omega + A \mid g \mid^2\right]g = 0. \tag{6.12}$$

As will be seen later the nature of the soliton solutions from (6.10) through (6.12) depend, in addition, on the boundary conditions applied to their solutions. The imaginary terms in both equations can be eliminated by taking the function $g(x,t)$ to be of the form of a uniformly translating envelope function, i.e., $g(x, t) = g_0(ax - bt)$. Then defining $y = ax{-}bt$ and choosing

$$b = v^2\frac{k}{\omega}a \tag{6.13a}$$

(6.11) is reduced to

$$\left[a^2 - \frac{b^2}{v^2}\right]\frac{\partial^2 g_0}{\partial y^2} + \left[-k^2 + \frac{\omega^2}{v^2} + A + B \mid g_0 \mid^2\right]g_0 = 0. \tag{6.13b}$$

Similarly, defining $y = ax{-}bt$ and choosing

$$b = \frac{\hbar k}{m}a \tag{6.14a}$$

(6.12) reduces to

$$-\frac{\hbar^2 a^2}{2m}\frac{\partial^2 g_0}{\partial y^2} + \left[\frac{\hbar^2 k^2}{2m} - \hbar\omega + A \mid g_0 \mid^2\right]g_0 = 0. \tag{6.14b}$$

Under the conditions stated in going from (6.10) to (6.14), the general form of the wave function solutions of (6.5) and (6.6) is given by

$$\psi(x,\, t) = \mathrm{e}^{\mathrm{i}(kx-\omega t)} g_0(ax - bt) \tag{6.15}$$

where $g_0(y)$ is from (6.13) or (6.14) solved with appropriate boundary conditions. The wave function in (6.15) is seen to be composed of a plane wave of phase velocity $v_\mathrm{p} = \frac{\omega}{k}$ which is modulated by an envelope function traveling along the x-axis with an envelope velocity $v_\mathrm{e} = \frac{b}{a}$. The velocities of the two features of the wave function are in general different.

Equations (6.13b) and (6.14b) are of the same general form, which may be written as

$$R\frac{\partial^2 g_0}{\partial y^2} + [S + T \mid g_0 \mid^2] g_0 = 0 \tag{6.16}$$

for appropriately defined real constants R, S, and T. For convenience R, S, and T for both the nonlinear Klein–Gordon and Schrödinger equations are summarized in table 6.1. The total soliton solutions for each of the systems (i.e., Klein–Gordon and Schrödinger) are seen to be obtained from (6.15) once a solution of g_0 is determined from (6.16).

Equation (6.16) is integrated using standard methods. Depending on the choice of boundary conditions, two types of soliton-like solutions are found. These have the spatial form of a kink which twists the envelope function through a sign change as x changes from negative to positive infinity and of a pulse which has zero amplitude at $x \to \pm\infty$. Both the envelope and plane wave portions of these two wave function types move along the chain with velocities v_e and v_p, respectively, that are uniquely defined by their equation types. For the kink soliton the envelope function is given in terms of a tanh function and has the general form

$$g_0(y) = W \tanh(y). \tag{6.17a}$$

The pulse soliton has an envelope function given in terms of a sech function and is of the general form

$$g_0(y) = W\ \mathrm{sech}(y). \tag{6.17b}$$

In these two expressions W is a real constant which is chosen to satisfy (6.16).

Table 6.1. The parameters R,S,T in (6.16) for the nonlinear Schrödinger (NS) equation and the nonlinear Klein–Gordon (NKG) equation [2].

Equation type	R	S	T
NS	$-\frac{\hbar^2 a^2}{2m}$	$\frac{\hbar^2 k^2}{2m} - \hbar\omega$	A
NKG	$a^2 - \frac{b^2}{v^2}$	$-k^2 + \frac{\omega^2}{v^2} + A$	B

Substituting the tanh form in (6.17a) into (6.16) gives

$$S = 2R \tag{6.18a}$$

and

$$W = \pm\sqrt{-\frac{S}{T}} \tag{6.18b}$$

for (6.16) to be satisfied. Using these equations and the results in table 6.1, a^2, W, and v_e are expressed in terms of (k,ω) and the basic parameters of the Klein–Gordon and Schrödinger equations. These results are listed in table 6.2.

In addition to the tanh solutions of the nonlinear (6.16) there are sech pulse solutions. To determine the values of a^2, W, and v_e which characterize the sech solution, substitution of the sech form in (6.17b) into (6.16) gives

$$R = -S \tag{6.19a}$$

and

$$W = \pm\sqrt{-\frac{2S}{T}}. \tag{6.19b}$$

Using these equations and the results in table 6.1, a^2, W, and v_e for the Klein–Gordon and Schrödinger systems are presented in table 6.3. These are expressed in terms of the basic constants of the Klein–Gordon and Schrödinger equations and (k, ω).

In the solutions for the solitons given in tables 6.2 and 6.3, the amplitudes, W, and the width, $2/|a|$, of the tanh and sech solitons depend on the envelope and phase

Table 6.2. Parameters for the tanh-soliton solution in (6.16) in terms of real k, ω, A, m, $\hbar$, and B [2].

Equation type	a^2	W	$v_e = \frac{b}{a}$
NS	$-\frac{1}{2}\frac{2m}{\hbar^2}\left[\frac{\hbar^2 k^2}{2m} - \hbar\omega\right]$	$\pm\sqrt{-\frac{\frac{\hbar^2 k^2}{2m} - \hbar\omega}{A}}$	$\frac{\hbar k}{m}$
NKG	$\frac{\frac{\omega^2}{v^2} + A - k^2}{2\left(1 - \frac{v^2 k^2}{\omega^2}\right)}$	$\pm\sqrt{-\frac{-k^2 + \frac{\omega^2}{v^2} + A}{B}}$	$v^2\frac{k}{\omega}$

Table 6.3. Parameters for the sech-soliton solution in (6.16) in terms of real k, ω, A, m, $\hbar$, and B [2].

Equation type	a^2	W	$v_e = \frac{b}{a}$
NS	$\frac{2m}{\hbar^2}\left(\frac{\hbar^2 k^2}{2m} - \hbar\omega\right)$	$\pm\sqrt{-2\frac{\frac{\hbar^2 k^2}{2m} - \hbar\omega}{A}}$	$\frac{\hbar k}{m}$
NKG	$\frac{k^2 - \frac{\omega^2}{v^2} - A}{1 - \frac{v^2 k^2}{\omega^2}}$	$\pm\sqrt{-2\frac{-k^2 + \frac{\omega^2}{v^2} + A}{B}}$	$v^2\frac{k}{\omega}$

velocities and do not vary as a function of time. Here $2/|a|$ characterizes the length over which the sech pulse is non-zero along the x-axis. For the tanh kink soliton $2/|a|$ characterizes the length of the twist region of the soliton along the x-axis. The W relates the maximum amplitude of either the tanh or sech type solitons. From the envelope velocities listed in tables 6.2 and 6.3, it is noted that the phase velocity and envelope velocities are always parallel for the soliton solutions of the nonlinear Klein–Gordon equation. For the nonlinear Schrödinger equation solutions the phase and envelope velocities are parallel to one another for soliton solutions with positive mass and anti-parallel for nonlinear Schrödinger equations with negative mass.

In addition to the earlier discussed soliton solutions that have envelopes which translate along the x-axis with a constant velocity, there are also modes that have stationary envelopes. These are known as breathers and they are of the form

$$\psi(x, t) = e^{-i\omega t} g(x). \tag{6.20}$$

Here the envelope function, $g(x)$, only depends on x and the envelope function is modulated by an harmonic, $e^{-i\omega t}$, time dependence.

Substituting (6.20) into (6.5) and (6.6) one finds an equation for $g(x) = g_0(y)$ where in the considerations of a stationary envelope $y = ax$. Making this substitution results in the equation

$$R\frac{\partial^2 g_0}{\partial y^2} + [S + T\,|\,g_0\,|^2]g_0 = 0. \tag{6.21}$$

Here the real constants R, S and T are listed in table 6.4 for both the Klein–Gordon and Schrödinger equations. The $g(x)$ solutions are of the form

$$g_0(y) = W\,\tanh(y) \tag{6.22a}$$

or

$$g_0(y) = W\,\text{sech}(y). \tag{6.22b}$$

Using these equations and the results in table 6.4, a^2, W, and v_e for the Klein–Gordon and Schrödinger systems are presented in tables 6.5 and 6.6. These are expressed in terms of the basic constants of the Klein–Gordon and Schrödinger equations and ω.

Table 6.4. The parameters R, S, T in (6.21) for the nonlinear Schrödinger equation and the nonlinear Klein–Gordon equation [2].

Equation type	R	S	T
NS	$-\frac{\hbar^2 a^2}{2m}$	$-\hbar\omega$	A
NKG	a^2	$\frac{\omega^2}{v^2} + A$	B

Table 6.5. Parameters for the tanh-soliton solution in (6.22a) in terms of real ω, A, m, $\hbar$, and$B$ [2].

Equation type	a^2	W	$v_e = \frac{b}{a}$
NS	$\frac{1}{2}\frac{2m}{\hbar^2}\hbar\omega$	$\pm\sqrt{\frac{\hbar\omega}{A}}$	0
NKG	$\frac{1}{2}\left(\frac{\omega^2}{v^2}+A\right)$	$\pm\sqrt{-\frac{\frac{\omega^2}{v^2}+A}{B}}$	0

Table 6.6. Parameters for the sech-soliton solution in (6.22b) in terms of real ω, A, m, $\hbar$, and B [2].

Equation type	a^2	W	$v_e = \frac{b}{a}$
NS	$-\frac{2m}{\hbar^2}\hbar\omega$	$\pm\sqrt{2\frac{\hbar\omega}{A}}$	0
NKG	$-\frac{\omega^2}{v^2} - A$	$\pm\sqrt{-2\frac{\frac{\omega^2}{v^2}+A}{B}}$	0

Table 6.7. The parameters R, S, T in (6.24) for the nonlinear Schrödinger equation and the nonlinear Klein–Gordon equation [2].

Equation type	R	S	T
NS	$-\frac{\hbar^2 a^2}{2m}$	$\frac{\hbar^2 k^2}{2m}$	A
NKG	$a^2 - \frac{b^2}{v^2}$	$-k^2 + A$	B

Another limit of interest is composed of a stationary plane wave along with an envelope which translates along the x-axis with a constant velocity. These solutions are of the form

$$\psi(x, t) = e^{ikx}g(x, t). \tag{6.23}$$

Substituting (6.23) into (6.5) and (6.6) one finds an equation for $g(x) = g_0(y)$ where $y = ax - bt$. This is given by

$$R\frac{\partial^2 g_0}{\partial y^2} + [S + T \mid g_0 \mid^2]g_0 = 0. \tag{6.24}$$

Here the real constants R,S, and T are listed in table 6.7 for both the Schrödinger and Klein–Gordon equations. The $g(x,t)$ solutions are of the form

$$g_0(y) = W\ \tanh(y) \tag{6.25a}$$

or

$$g_0(y) = W\ \mathrm{sech}(y). \tag{6.25b}$$

From these forms it is found that for the nonlinear Schrödinger equation

$$v_e = \frac{b}{a} = \frac{\hbar k}{m} \tag{6.26}$$

is the velocity of the envelope. For the nonlinear Klein–Gordon equation the condition on the solutions is that

$$k = 0. \tag{6.27}$$

Consequently, for the Klein–Gordon equation the translating pulse has no plane wave but only an envelope function. The further details of the solution of this case are left to be worked out by the reader.

References

[1] Dauxois T and Peyrard M 2006 *Physics of Solitons* (Cambridge: Cambridge University Press)

[2] Giri P, Choudhary K, Sen Gupta A, Bandyopadhyay A K and McGurn A R 2011 Klein–Gordon equation approach to nonlinear split-ring resonator based metamaterials: one-dimensional systems *Phys. Rev.* B **84** 155429

Arthur R McGurn

Chapter 7

Difference equations in physics

Many of the dynamical equations for the excitations in materials are of the form of difference equations [1–12]. These arise using a basis which is defined locally about the sites of the crystal lattice of the system to rewrite its dynamical equations in a matrix format. In photonic crystals the basis is defined in the unit cells of the dielectric lattice while in metamaterials it is associated with the individual split ring resonator (SRR) units. In the study of vibrational properties of crystals the basis is the individual cluster of atoms in the unit cell, while for the electronic properties of graphene the basis is the unit cell of carbon atoms [13]. In all these varied systems the dynamics is written in terms of the interactions of the basis units, and the resulting algebraic eigenvalue problems are particular expressions of the difference equation format. In the following we look at the difference equation approach with a focus on applying it to study the excitations in photonic crystals and metamaterials. An additional focus is on the effects of nonlinearity on the difference equation solutions and the new types of excitations that arise in the system in the presence of nonlinearity. To simplify the discussions the presentation will be made for one-dimensional systems, though modifications to handle three-dimensional cases are straightforward. At the end some contact will be made with the continuum limit (differential equation) approach to the dynamics of these systems.

Well known examples of difference equations for one-dimensional systems defined along an x-axis are [1–8, 13]:

a. The dynamical equations for the vibrational properties of a linear chain of atoms:

$$-m\omega^2 x_n + k_h [2x_n - x_{n+1} - x_{n-1}] = 0 \tag{7.1}$$

where m is the atom mass, x_l is the displacement of the lth atom along the x-axis, k_h is the spring constant, and ω is the mode frequency. This is treated in most condensed matter physics texts.

doi:10.1088/978-0-7503-3579-9ch7

b. The dynamical equations for boson or electron excitations in the approximation of the tight binding model:

$$-\hbar\omega c_n^+ + h_0 c_n^+ + h_1\,[c_{n+1}^+ + c_{n-1}^+] = 0 \tag{7.2}$$

where c_l^+ is a Bose (Fermi or C-number) operator at the lth lattice site, h_0 and h_1 are coupling constants, and ω is the mode frequency.

c. The dynamical equations of the guided modes of a photonic crystal waveguide:

$$E_n - g\alpha E_n - g\beta\,[E_{n+1} + E_{n-1}] = 0 \tag{7.3}$$

where E_l is the electric field at the lth site, g is related to the difference in the dielectric properties of the waveguide channel from the photonic crystal, and α and β are couplings which depend on the frequency ω of the electromagnetic mode.

d. The dynamical equations of the modes of an SRR or metamaterial chain:

$$-\omega^2\left[q_n + \lambda(q_{n+1} + q_{n-1})\right] + q_n = 0 \tag{7.4}$$

where q_l measures the charge on the split gap of the SRR at site l, λ is a measure of the ratio of the strength of the mutual inductive coupling between nearest neighbor SRRs to the self-inductance of the individual SRR.

All of these models are linear difference equations. They all share the common form:

$$A(\omega)s_n + B(\omega)\,[s_{n+1} + s_{n-1}] = 0 \tag{7.5}$$

for coefficients $A(\omega)$ and $B(\omega)$ and the dynamical variables s_l.

Due to the translational symmetry along the chain, the modes of (7.5) are plane waves of the form

$$s_l = s\mathrm{e}^{ikl} \tag{7.6a}$$

where k is the wave number of the plane wave and l labels the chain sites. Upon substituting (7.6a) in (7.5) the dispersion relation of the mode frequency ω is the solution of

$$A(\omega) + 2B\,(\omega)\cos k = 0. \tag{7.6b}$$

For the case of the harmonic coupled chain of atoms the application of the theory of (7.5) through (7.6b) results in the well know dispersion relation for the linear chain with a monatomic basis

$$m\omega^2 = 2k_h\,(1 - \cos k) \tag{7.7a}$$

and from (7.4) a similar substitution for the q_l charges of the SRR metamaterial yields the dispersion relation

$$\omega^2 = 1/\,[1{+}2\lambda\cos k]. \tag{7.7b}$$

Similar types of results are found for the other previously listed systems.

7.1 Nonlinear photonic crystals and SRR metamaterials

Kerr nonlinearity is often present in photonic crystal and metamaterial optical systems [1–7]. In these systems the difference equations of (7.5) are modified to take the general form

$$A(\omega)s_n + B(\omega)[s_{n+1} + s_{n-1}] + C(\omega) \mid s_n \mid^2 s_n + D(\omega)[\mid s_{n+1} \mid^2 s_{n+1} + \mid s_{n-1} \mid^2 s_{n-1}] = 0 \quad (7.8)$$

where the nonlinear terms in $C(\omega)$ and $D(\omega)$ are to be treated as perturbations. In optical systems nonlinearity is generally a small effect and has traditionally become an area of study only with the advent of lasers. A problem of modern material science is the development of materials with enhanced nonlinear optical effects. In certain instances it is possible to generate larger fields within the nonlinear media of optical devices by manipulating the geometry and materials in their design and this can aid in the observation of effects arising from nonlinearity.

The nonlinear terms in (7.8) contain absolute values of the variables s_j so that the solutions of (7.8) only involve a single frequency $e^{\pm i\omega t}$. As a consequence, higher harmonics do not arise from adding nonlinearity to the originally linear equations. This follows as the equations have been generated in an approximation known as the rotating wave approximation (RWA). In the RWA the gains and losses of the fundamental ($e^{i\omega t}$-dependent terms of the system) arising through the generation and absorption of higher harmonics within the nonlinear system are ignored. As an example, without the RWA approximation, the third order nonlinearities in the photonic crystal and SRR metamaterial involve powers in $s_j = se^{i\omega t} + cc$ of the form

$$s_j^3 = s^3 e^{i3\omega t} + 3 \mid s \mid^2 se^{i\omega t} + 3 \mid s \mid^2 s_* e^{-i\omega t} + (s_*)^3 e^{-i3\omega t}. \quad (7.9)$$

Retaining only the $e^{i\omega t}$ terms in (7.8) and (7.9) decouples the equation for the fundamental frequency from those for the frequencies e^{i0t}, $e^{-i\omega t}$, $e^{\pm i3\omega t}$, etc, and their interactions with one another as generated by the s_j^3 nonlinearities. The resulting single frequency equation in (7.8) comes from this decoupling and is a representation of the system for a Kerr medium described by the Kerr interaction.

The nonlinear terms in (7.8) lend to a variety of interesting new effects [1, 9]. The dispersion relations of the extended plane wave solutions of (7.8) now depend, through the $\mid s \mid^2$ interactive terms, on the amplitude of the waves. In addition, new types of excitations which are not present in the linear system appear. These excitations include soliton modes of the bright, dark, and gray soliton types. Bright soliton modes are intensity or field pulses that propagate at constant velocities along the chain. These pulses can be modulated by a propagating phase. Dark solitons are a complete dip or absence in the intensity or field that propagates through the system at constant velocities, and again may be modulated by a propagating phase. Gray solitons are similar to dark solitons. They are minima but not total absences in the intensity or field propagating at constant velocities which may modulate a propagating phase.

The origins of the new soliton solutions are the interaction of the mode with the nonlinear media and the dependence of the nonlinear medium properties on the mode amplitude and shape [9]. Tuning the mode amplitude and shape along the chain allows the system to change its dielectric properties so as to support the intensity peak or dip of the mode. The mode affects the medium to self-consistently support itself. This requires that a precise set of conditions be satisfied on the mode shape, amplitude, and velocity. An additional feature of these new excitations and the modified plane wave solutions is that they occur at different frequencies and energies. Depending on their relative energies in the nonlinear system, a higher energy mode may breakup or decay into modes of lesser energies that are more stable in the system. This has consequences for the lifetimes and stability of the new soliton solutions in the system.

For the following discussion of the solutions of the discrete systems of (7.8), the focus will be on soliton-like modes that have stationary envelopes. This is a simplification in the presentation, and the reader is referred to the literature for studies of the case of modes involving constant velocity envelope motion. Constant velocity envelope motion can be handled both analytically and numerically in a manner which involves generalization of the methods discussed later. The stationary envelope solutions that we discuss here are generally known as discrete breathers because their solutions tend to involve a stationary envelope which is modulated by a periodic time variation. These types of modes are of interest as they are the primary type of modes confined to nonlinear devices that form part of an optical circuit element. It is the properties of the stationary soliton modes in these components which influence the function of the circuit.

7.2 Bright soliton-like solutions

A solution of the nonlinear equation in (7.8) for a symmetric bright soliton-like pulse centered about the site index $n = 0$ can be obtained [1, 11] as a linear combination of the basis set $\{| n |^{m-1} A_m e^{-(|n|-1)r}\}$. In this set of functions the sites of the chain are labeled by the site index, n, running from negative to positive infinity, and m is a summation index used in forming linear combinations of the linearly independent basis elements. In this basis one form of solution for the $\{s_n\}$ is written as

$$s_0 = \alpha_0 \tag{7.10a}$$

and for $n \neq 0$

$$s_n = \alpha_0 \left[\sum_{m=1}^{\infty} A_m |n|^{m-1} e^{-(|n|-1)r} \right]. \tag{7.10b}$$

Here α_0 is the peak height of the pulse and (7.10b) gives the envelope which decays away from the peak height. Another second solution with a bright pulse form is given as

$$s_0 = \alpha_0 \tag{7.11a}$$

and for $n \neq 0$

$$s_n = \alpha_0 \left[\sum_{m=1}^{\infty} A_m (-1)^n |n|^{m-1} e^{-(|n|-1)r} \right] \tag{7.11b}$$

where, again, the peak height is α_0 and (7.11b) is the decaying envelope of the wave amplitude. These forms differ fundamentally. The amplitude of the first form in (7.10) steadily or continuously decays from its maximum peak height as one moves along the chain, whereas the amplitude of the form in (7.11) oscillates in sign from site to neighboring site as its envelope decays from its maximum peak height.

The number of terms required in the sum over m in (7.10b) and (7.11b) depends on the width of the pulse being studied, for as the pulse broadens in width more and more terms of different m are needed to accurately represent it. As an example of this consider a narrow pulse. For sharply peaked pulses only the $m = 1$ terms need to be retained in (7.10b) and (7.11b). With this simplification (7.10) reduces to

$$s_0 = \alpha_0 \tag{7.12a}$$

and for $n \neq 0$

$$s_n = \alpha_0 A_1 e^{-(|n|-1)r}. \tag{7.12b}$$

Likewise, (7.11) reduces to

$$s_0 = \alpha_0 \tag{7.13a}$$

and for $n \neq 0$

$$s_n = \alpha_0 A_1 (-1)^n e^{-(|n|-1)r}. \tag{7.13b}$$

Substituting (7.12) into the set of difference equations in (7.8) for n=0, 1, and $n \to \infty$ gives three nonlinear equations to be solved for A_1, e^{-r}, and ω as functions of the peak height α_0 and the parameters other than ω characterizing $A(\omega)$ and $B(\omega)$. Making these substitutions generates the following set of three nonlinear equations for the three unknowns

$$A(\omega) + 2B(\omega)A_1 + C(\omega)\alpha_0^2 + 2D(\omega)\alpha_0^2 A_1^3 = 0 \tag{7.14a}$$

$$A(\omega)A_1 + B(\omega)[A_1 e^{-r} + 1] + C(\omega)\alpha_0^2 A_1^3 + D(\omega)\alpha_0^2 [A_1^3 e^{-3r} + 1] = 0 \tag{7.14b}$$

$$A(\omega) + 2B(\omega)\cosh(r) = 0. \tag{7.14c}$$

Similarly, substituting (7.13) into the difference equations in (7.8) for $n = 0$, 1, and $n \to \infty$ gives three nonlinear equations to be solved for A_1, e^{-r}, and ω as functions of the peak height α_0 and the parameters other than ω characterizing $A(\omega)$ and $B(\omega)$. Making the substitutions generates the following set of three nonlinear equations for the three unknowns

$$A(\omega) - 2B(\omega)A_1 + C(\omega)\alpha_0^2 - 2D(\omega)\alpha_0^2 A_1^3 = 0 \tag{7.15a}$$

$$A(\omega)A_1 - B(\omega)[A_1 e^{-r} + 1] + C(\omega)\alpha_0^2 A_1^3 - D(\omega)\alpha_0^2 [A_1^3 e^{-3r} + 1] = 0 \quad (7.15b)$$

$$A(\omega) - 2B(\omega)\cosh(r) = 0. \quad (7.15c)$$

To illustrate the numerical solutions of the case of a narrow peak, the solutions of (7.14) and (7.15) are next studied for some particular nano-systems involving photonic crystal and SRR metamaterial waveguides.

In the presence of Kerr nonlinearity the difference equations, obtained within the RWA, of the metamaterial system from (5.5), (7.4), (7.8), and (7.9) take the form

$$-\omega^2\left[q_n + \lambda(q_{n+1} + q_{n-1})\right] + q_n - \frac{3\alpha}{\varepsilon_l} | q_n |^2 q_n = 0. \quad (7.16)$$

Here the last term on the left-hand side comes from the Kerr nonlinearity, ε_l is the linear term of the Kerr dielectric, and α is the sign of the mutual inductive couplings. Evaluating the narrow peak limit equations in (7.14) and (7.15) for the parameters of the (7.16) form generates the following system of nonlinear equations [1, 11]

$$\frac{3\alpha}{\varepsilon_l}\alpha_0^2 \pm 2\lambda A_1 \mp 2\lambda\left[1 - \frac{3\alpha}{\varepsilon_l}\alpha_0^2\right]\cosh(r) = 0 \quad (7.17a)$$

$$\lambda(1 + A_1 e^{-r}) \pm \frac{3\alpha}{\varepsilon_l}\alpha_0^2 A_1^3 - 2\lambda\left[A_1 - \frac{3\alpha}{\varepsilon_l}\alpha_0^2 A_1^3\right]\cosh(r) = 0 \quad (7.17b)$$

and

$$\omega^2 = [1 \pm 2\lambda\cosh(r)]^{-1}. \quad (7.17c)$$

In these equations, the upper (lower) signs are for the (7.12) ((7.13)) forms of the pulse discussed earlier.

In table 7.1 results from a set of published solutions of (7.17) are presented for a parameterization of the metamaterial that has recently been studied in [1] and [11]. The table presents bright soliton results for a number of parameter sets (λ, $\frac{\alpha}{\varepsilon_l}$, α_0). It is seen that the values λ are consistent with weak nearest neighbor couplings in the system, and $\frac{\alpha}{\varepsilon_l}$ and the peak amplitude α_0 set the dielectric and amplitude conditions needed to observe a bright soliton solution of frequency ω. The last three columns list the amplitude solutions of the pulse at the lattice sites $n = 0, 1, 2$. For the results presented, the solutions of the bright pulses are peaked at $n = 0$ and the amplitudes

Table 7.1. Bright stationary breather solutions of (7.17) for recently studied metamaterial system [11].

λ	α/ε_l	α_0	ω	q_0	q_1	q_2
0.0165	0.00012	5.640	0.9816	5.640	4.451	2.594
−0.0165	0.00012	5.640	0.9816	5.640	−4.451	2.594
0.0175	0.00028	2.880	0.9817	2.880	2.497	1.706
−0.0175	0.00028	2.880	0.9817	2.880	−2.497	1.706

of the pulses decay in absolute magnitude as one proceeds away from $n = 0$ and along the chain. For one set of pulses the amplitudes are all positive along the chain, whereas for the other set of pulses the amplitudes oscillate in sign as one moves along the chain, and for each set of chain parameters the frequencies of these two types of pulses are numerically the same. As per the initial assumptions on our system, the solutions exhibit a rapid decay in magnitude from the peak values. In both cases, due to the weak intra-chain couplings λ, the dispersion relations of the pulse modes are little changed from the dispersion relation of the plane waves in the chain for $\lambda = 0$. It is found in general that the larger the peak heights of the excitations the larger are the nonlinear effects displayed by the system solutions, the narrower are the widths of the peaks, and the greater the excitation frequencies differ from the $\lambda = 0$ limit.

In the presence of Kerr media, the photonic crystal waveguide in (7.3) for a channel composed of linear media is modified to read (see also (5.13))

$$E_n - g\alpha(\omega)[1 + \lambda\,|E_n|^2]E_n - g\beta(\omega)\{[1 + \lambda\,|E_{n+1}|^2]E_{n+1} + [1 + \lambda\,|E_{n-1}|^2]E_{n-1}\} = 0, \quad (7.18)$$

where λ is the Kerr parameter of the Kerr nonlinear dielectric, and for $\lambda \to 0$ the system reverts to that of the linear waveguide of (7.3). In (7.18) g is proportional to the difference in the linear part of the dielectric constants between the dielectric cylinders of the photonic crystal and their replacements in the waveguide channel, $\alpha(\omega)$ and $\beta(\omega)$ are frequency-dependent couplings that are expressed in the Green's functions of the bulk photonic crystal, and the frequency ω is set within a stop band of the bulk photonic crystal. Due to the frequency dependence of $\alpha(\omega)$ and $\beta(\omega)$ it is best to treat (7.18), similarly to (3.32), as an eigenvalue problem for the eigenvalue g. Equation (7.18) then determines the dielectric contrast g needed to support a guided mode of frequency ω in the waveguide, and the eigenvectors are the amplitudes of the modes in the system. In the following, the discussions will follow those given for the soliton modes of the SRR metamaterials, i.e., the eigenvectors will be assumed to be of the form of pulses characterized by parameters which are to be determined. The pulse form is substituted into (7.18) generating equations for the pulse parameters and the eigenvalue g.

As an illustration of the procedure for studying the bright soliton solutions of (7.18), the narrow pulse equations will be obtained. The treatment is essentially the same as that made for the SRR metamaterial in (7.16) through (7.17). Taking a pulse form similar to that in (7.13)

$$E_0 = \alpha_0 \quad (7.19a)$$

and for $n \neq 0$

$$E_n = \alpha_0 A_0 (-1)^n e^{-|n|r} \quad (7.19b)$$

and substituting into (7.18), three nonlinear equations for the parameters of the pulse form and eigenvalue are obtained. For a pulse centered at $n = 0$ the three equations are generated by considering the same three sites as were considered in the case of the SRR metamaterial. These give [14]:

(a) for $n = 0$

$$1/\gamma_0 = 1 + a - 2b\Delta - 2ab\Delta^3 \tag{7.20a}$$

(b) for $n = 1$

$$1/\gamma_0 = \frac{1}{\Delta}\left[\Delta + a\Delta^3 - b\left(1 + \Delta K\right) - ab\left(1 + \Delta^3 K^3\right)\right] \tag{7.20b}$$

(c) for $n \to \infty$

$$1/\gamma_0 = 1 - 2b\cosh(r) \tag{7.20c}$$

where $\gamma_0 = g\alpha$, $\Delta = A_0 e^{-r}$, $a = \lambda\alpha_0^2$, $b = \beta/\alpha$, and $K = e^{-r}$. Equations (7.20) are solved for γ_0, Δ, and K in terms of the a,b. The parameter a involves the Kerr parameter which sets the strength of the nonlinearity and the maximum peak height of the field amplitude, while b is the ratio of the nearest neighbor coupling to the onsite coupling. The parameter b in our treatment needs to be small so that only nearest neighbor couplings are important in the difference equations. This is easily arranged as the wave function of the electric field, at guided wave frequencies in the stop band of the bulk photonic crystal, generally decays with distance rapidly in an exponential manner. By increasing the nearest neighbor separation of the cylinders forming the waveguide channel b quickly decreases. The wave function, obtained from (7.19), is then easily expressed in terms of the solutions for γ_0, Δ, and K from (7.20).

Analytic solutions of (7.20) can be obtained in the limit of highly localized pulses, and this limit is found to be helpful in later investigations of the nature of discrete breathers for possible device applications. In the case of a highly localized pulse for which the width determining parameters are very small, i.e., $|\Delta|$, $|K| \ll 1$, analytic forms have been generated for the solutions of (7.20). In terms of the a and b these are [15]

$$K = \frac{a \pm a\left[1 + 8\left(a + 1\right)b^2/a^2\right]^{1/2}}{4\left(a + 1\right)b} \tag{7.21a}$$

$$\Delta = \left(a + 1\right)K \tag{7.21b}$$

and

$$\gamma_0 = \left[1 - \frac{b}{K}\right]^{-1}. \tag{7.21c}$$

Equations (7.21) represent analytic forms giving highly localized pulse solutions of the eigenvalue problem but are found to be valid over a wide range of a and b. A comparison will now be made of the analytic and numerical results for (7.20).

The three nonlinear equations of (7.20) can be solved numerically using the Newton–Raphson method and compared with the limiting forms in (7.21). In figure 7.1 results are shown for the solution of (7.20) with a comparison of the

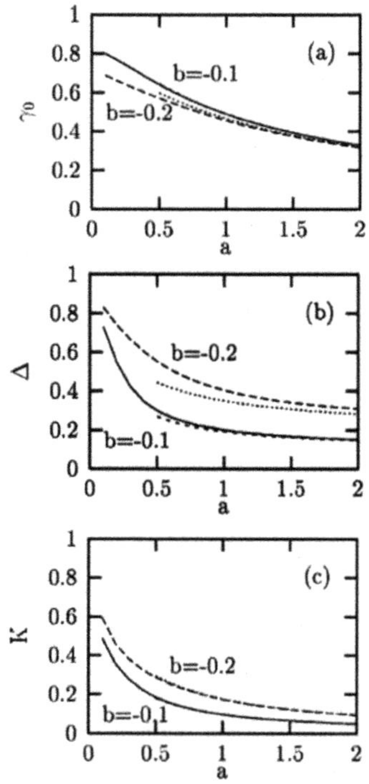

Figure 7.1. Solutions for: (a) γ_0, (b) Δ, and (c) K versus a for representative values of b. Reproduced with permission from [15]. Copyright 1999 American Institute of Physics.

limiting forms in (7.21) [14, 15]. The parameters a and b for the results in the plots were chosen to cover a wide range of systems of possible physical interest and are discussed in detail in [15]. Plots are presented for γ_0, Δ, and K versus a for typical small values of b. The dotted lines in figure 7.1(a) and (b) show corresponding results from the limiting forms in (7.21). In figure 7.1(c), the difference between the numerical solutions and the analytical forms cannot be resolved on the scale of the plot. These limiting forms are seen to be reasonably accurate over a wide range of a. As the nonlinearity and maximum peak height, indicated by a, are decreased for a fixed b as seen from figure 7.1(b) and (c) the pulse is observed to broaden as it begins to approach the limit of a linear system. This holds for both the parameter related to the exponential decay along the chain, $K = e^{-r}$, and for the parameter for the decay between nearest neighbor sites along the chain, $\Delta = A_0 e^{-r}$. In addition, for fixed a, the width related parameters are found to decrease with decreasing |b|. Likewise, from figure 7.1(a) the parameter, $\gamma_0 = g\alpha$, which is related to the dielectric contrast is found to increase with decreasing a for a fixed b and to decrease with increasing |b| for a fixed a.

Similar discussions to those just made for the photonic crystal system in (7.20) and (7.21) can be given for the SRR metamaterial systems described by (7.17). The results presented in table 7.1 for the SRR metamaterials are representative of isolated points from plots similar to those in figure 7.1 that can be generated for the

SRR metamaterials. Later it will be shown how both types of presentations for the soliton modes of a system can find practical applications in the discussion of device related systems.

Additional bright soliton pulse solutions have been studied in the photonic crystal system that are significantly different from those discussed earlier. An additional asymmetric pulse exists and has been shown to be possible in photonic crystals. To understand this new pulse, consider a one-dimensional lattice in which the site labeled $n = 0$ is deleted and the sites labeled $n = 1$ and $n = -1$ are considered to be nearest neighbors. This is just a renumbering of the lattice consecutively with the omission of 0. For this renumbering new trial solutions of the form of a pulse centered about the sites labeled $n=\pm 1$ is obtained by assuming a trial form given by [14, 15]

$$E_1 = -E_{-1} = \alpha_0 \tag{7.22a}$$

and for $n \neq 0$

$$E_n = -E_{-n} = \alpha_0 A_0 \left(-1\right)^{n-1} \mathrm{e}^{-(n-1)r} \tag{7.22b}$$

where $n>0$. Again an illustration of this pulse solution will be made by studying the solutions in the limit of a narrow pulse. In this limit, the solution procedure is much the same as that used in the studies of (7.17) and (7.20).

Upon substituting (7.22) into (7.18) for the sites labeled $n = 1$, 2, and $n\rightarrow\infty$ three self-consistent equations are found [14, 15]:

(a) for $n = 1$

$$1/\gamma_0 = 1 + a - b\left(\Delta + 1\right) - ab\left(\Delta^3 + 1\right) \tag{7.23a}$$

(b) for $n = 2$

$$1/\gamma_0 = \frac{1}{\Delta}\left[\Delta + a\Delta^3 - b\left(1 + \Delta K\right) - ab\left(1 + \Delta^3 K^3\right)\right] \tag{7.23b}$$

(c) for $n\rightarrow\infty$

$$1/\gamma_0 = 1 - 2b\cosh\left(r\right). \tag{7.23c}$$

Here $\gamma_0 = g\alpha$, $\Delta = A_0 \mathrm{e}^{-r}$, $a = \lambda\alpha_0^2$, $b = \beta/\alpha$, and $K = \mathrm{e}^{-r}$, and it is seen that the parameterization is the same as that given in (7.20). Again, (7.23) are solved for γ_0, Δ, and K in terms of the a, b. The parameter a involves the Kerr parameter which sets the strength of the nonlinearity and the maximum peak height of the field amplitude, while b is the ratio of the nearest neighbor coupling to the onsite coupling which needs to be small for a nearest neighbor coupling model. The wave function, obtained from (7.22), is then easily expressed in terms of the solutions for γ_0, Δ, and K obtained from (7.23).

As with the earlier photonic crystal solution, in the limit of a narrow bright solution pulse an analytic limiting form of solution can be given. In the case of a highly localized pulse for which the width parameters $|\Delta|$, $|K| \ll 1$, the solution of (7.23) is [14, 15]

$$K = \frac{a - b(1 - a) \pm [(a - b\,(1 - a))^2 + 4\,(a + 1)b^2]^{1/2}}{2(a + 1)b} \tag{7.24a}$$

$$\Delta = (a + 1)K \tag{7.24b}$$

and

$$\gamma_0 = \left[1 - \frac{b}{K}\right]^{-1}. \tag{7.24c}$$

Again the solutions of γ_0, Δ, and K obtained from these results can easily be used to obtain the bright solution wave functions in (7.22).

7.3 Dark and gray soliton solutions

Another set of solitons are the dark and gray solitons [1, 9, 11, 14, 15]. These are solutions of the nonlinear difference equations that have a uniform intensity over the length of the waveguide chain except for an intensity dip at one place along the chain. If the intensity of the dip goes to zero, the soliton is a dark soliton, but if the intensity of the dip does not go completely to zero, the soliton is a gray soliton. Both of these types of solitons will be now treated. To simplify the treatment, only the limit of a dip with a narrow width will be treated here. For the more general treatment of modes with general widths, which is very similar in the steps applied to that of the general treatment of the bright soliton, the reader is referred to the literature.

In the limit of a narrow intensity dip, a similar treatment to that given earlier for bright soliton modes can be made for dark and gray soliton modes. A trial solution form which has an intensity dip is substituted in the nonlinear difference equations and its parameters are set to approximately satisfy the general features of the dark or gray soliton solution over the length of the waveguide chain. In the following, the narrow dark and gray modes are investigated in the discrete model by assuming trial solutions of the form [11]

$$s_0 = \mathrm{i}d \tag{7.25a}$$

$$s_n = \{\alpha_0(1 - A_0\mathrm{e}^{-nr} - nA_1\mathrm{e}^{-nr}) + \mathrm{i}d\}\mathrm{e}^{\mathrm{i}n\delta} \tag{7.25b}$$

and

$$s_{-n} = -\{\alpha_0(1 - A_0\mathrm{e}^{-nr} - nA_1\mathrm{e}^{-nr}) - \mathrm{i}d\}\mathrm{e}^{-\mathrm{i}n\delta} \tag{7.25c}$$

for $n > 0$. Here $d = 0$, $\delta = 0$ generates dark solitons and $d \neq 0$, $\delta \neq 0$ generates gray solitons. In the case of the gray solitons $d \neq 0$ sets the asymmetric boundary conditions of the field amplitudes for the gray modes at $n \to \pm\infty$, while for dark solitons $d = 0$ assures that the intensity of the dark modes are the same at $n \to \pm\infty$. For $d = 0$, $\delta = 0$ the dark soliton form is seen to have a complete intensity dip at $s_0 = 0$ while the field amplitude approaches the limiting forms $s_{\pm n} \to \pm\alpha_0$ as $n \to \infty$ giving a constant nonzero intensity. The $d \neq 0$, $\delta \neq 0$ gray soliton has a nonzero dip at $s_0 = id$ and approaches a constant intensity as $n \to \infty$. For a generalization of (7.25) to treat modes of general width [11], it is noted in the literature that the set $\{n^m A_m e^{-nr}\}$ for $m > 0$ is a linearly independent basis that can be used to write the general dark and gray solutions of the set of difference equations considered in the present discussions. This general case is not treated here.

For the narrow dip mode solutions, the set of trial forms in (7.25) are sufficient to describe intensity dip modes that quickly approach their constant amplitude $n \to \infty$ limits. As with narrow peaked bright solitons only a few of the difference equations need be satisfied in order to generate a reasonable approximation to the dark and gray modes. In the following the dark soliton solutions in the metamaterial system of (7.16) will be treated. As with the bright soliton considerations in (7.17) the trial form in (7.25) is used to satisfy the difference equations for $n = 0, 1, 2$, and $n \to \infty$.

For dark solitons $d = 0$ and $\delta = 0$, and a solution can be obtained from (7.25) taking $A_1 = 0$. In the following we consider dark soliton solutions of the form of (7.25) and also an additional dark soliton mode closely related to that of (7.25) for the metamaterial system in (7.16). Using the notation of (7.16) the first form, obtained directly from (7.25), is given by [11]

$$q_0 = 0 \tag{7.26a}$$

$$q_n = \alpha_0(1 - A_0 e^{-nr}) = -q_{-n} \tag{7.26b}$$

and the second form, obtained as a small modification of (7.25), is

$$q_0 = 0 \tag{7.27a}$$

$$q_n = \alpha_0(-1)^n(1 - A_0 e^{-nr}) = -q_{-n}. \tag{7.27b}$$

Note that the first solution and the second solution only differ by an oscillating sign as one passes along the waveguide chain of sites.

Substitution of the two separate forms in (7.26) and (7.27) into the metamaterial (7.16) for $n = 1, 2$ and $n \to \infty$ generates three nonlinear equations [11]

$$-\omega^2\,[1 \pm 2\lambda] + 1 - \frac{3\alpha}{\varepsilon_l}\alpha_0^2 = 0 \tag{7.28a}$$

$$-\omega^2\left\{1 - A_0 e^{-r} \pm \lambda[1 - A_0 e^{-2r}]\right\} + \left\{1 - \frac{3\alpha}{\varepsilon_l}\alpha_0^2(1 - A_0 e^{-r})^2\right\}(1 - A_0 e^{-r}) = 0, \tag{7.28b}$$

$$\begin{aligned} &-\omega^2\left\{1 - A_0 e^{-2r} \pm \lambda\,[2 - A_0(e^{-r} + e^{-3r})]\right\} \\ &+\left\{1 - \frac{3\alpha}{\varepsilon_l}\alpha_0^2(1 - A_0 e^{-2r})^2\right\}(1 - A_0 e^{-2r}) = 0 \end{aligned} \tag{7.28c}$$

for A_0, e^{-r}, and ω^2. The upper (lower) signs are for the first (second) form of solutions. Similar to the treatment of the bright soliton equations, specifying the values of the set of parameters $\left(\lambda, \frac{\alpha}{\varepsilon_l}, \alpha_0\right)$, these equations are solved for A_0, e^{-r}, and ω^2. This gives a complete determination of the wave function and frequency of the dark soliton mode in terms of the inductive couplings, dielectric constants and the magnitude of the wave function amplitude at $n \to \pm\infty$. Note that the rate of healing of the dip of the mode is most influenced by the factor of e^{-r}. This is seen from the (7.26b) and (7.27b) forms.

In the narrow mode limit, analytic forms can be given for the solutions of (7.28). In the limit that $|e^{-r}| \ll 1$ and $\frac{3\alpha\alpha_0^2}{\lambda\varepsilon_l}e^{-r}$ is of order 1, limiting forms for the solution of the above equations are

$$A_0 \cong 1 \tag{7.29a}$$

$$e^{-r} \cong \mp \frac{1}{2\frac{3\alpha\alpha_0^2}{\lambda\varepsilon_l}} \tag{7.29b}$$

and

$$\omega^2 \cong 1 \mp \lambda\left[2 \pm \frac{3\alpha\alpha_0^2}{\lambda\varepsilon_l}\right]. \tag{7.29c}$$

Again these expressions are technically only valid for very narrow pulses, but, like the bright soliton limiting forms, they seem to work semi-quantitatively outside these limits.

In table 7.2 results are presented from a numerical evaluation of (7.28) that is found in the literature [1–7, 11]. The values of λ are chosen to be small for a weak nearest neighbor coupling along the waveguide chain and are similar to those found in a number of recent studies of such systems. The parameters $\frac{\alpha}{\varepsilon_l}$ are then fixed by the conditions for the existence of a dark soliton with amplitude $|q_3| \approx |q_{\pm\infty}| = \alpha_0$. The amplitude α_0 is a chosen parameter of the computation. (Note that the choice of α_0 used in table 7.2 was set by arguments, not given here, that were made in a recent publication and these arguments are not important to our purposes here.) As with the bright solitons, the dark solitons have frequencies, ω, close to those of the λ=0 plane wave modes that are given by (7.7b). The last four columns of the table give the field amplitudes near the center of the dark soliton dip at $q_0 = 0$. The dip rapidly heals as one leaves the core of the soliton dip to the amplitude magnitudes at infinity.

Table 7.2. Dark soliton solutions [11].

λ	α/ε_l	ω	q_0	q_1	q_2	q_3
0.0255	−0.00260	0.9792	0.0000	0.3677	0.6382	0.7929
0.0270	−0.00332	0.9789	0.0000	0.4068	0.6883	0.8362
0.0295	−0.00400	0.9776	0.0000	0.4277	0.7133	0.8564
0.0315	−0.00500	0.9772	0.0000	0.4624	0.7515	0.8853

Some of these results for the gray solitons are plotted later in figure 8.5 where they are discussed within another context.

A similar discussion to that given for SRR metamaterials can be made for the dark soliton modes in photonic crystal systems. Here the dark soliton photonic crystal modes will be illustrated by discussions of their solutions in the narrow mode limit. In this limit, the trial wave function of the dark soliton mode in photonic crystal waveguides is made by analogy with the discussions for the SRR metamaterial waveguides. The field amplitudes along the waveguide chain are of the form [14, 15]

$$E_0 = 0 \tag{7.30a}$$

$$E_n = \alpha_0(-1)^n(1 - A_0 e^{-nr}) = -E_{-n}. \tag{7.30b}$$

Substituting (7.30) into (7.18) and evaluating the difference equations for $n = 0, 1, 2,$ and $n \to \infty$ generates a number of nonlinear equations for the unknowns $\gamma_0 = g\alpha$, A_0, and e^{-r}. The unknowns are solved for in terms of the parameters $\gamma_0 = g\alpha$, $a = \lambda\alpha_0^2$, $b = \beta/\alpha$, and e^{-r}, obtained from the coefficients of the difference equation. An analytic solution of the resulting nonlinear equations in the limit that $|\frac{\beta}{\alpha}|, |\frac{\beta}{\alpha}\lambda\alpha_0^2| \ll 1$ is

$$\gamma_0 = [1 + a - 2b - 2c]^{-1} \tag{7.31a}$$

$$A_0 = (b + c)/(b + 3c) \tag{7.31b}$$

and

$$e^{-r} = (b + 3c)/(2a). \tag{7.31c}$$

Here $a = \lambda\alpha_0^2 \ll 1$, $b = \frac{\beta}{\alpha} \ll 1$, $c = ab$, and $|e^{-r}| \ll 1$. The solutions are in the form of a narrow intensity dip which rapidly heals to the constant amplitude at infinity.

7.4 Continuum limits

In chapters 5 and 6 the continuum limits of certain difference equations were treated along with a discussion of the soliton solutions of the resulting differential equations. The study of solitons was actually first made in terms of differential equations so that it is natural in a treatment of these excitations to first review bright and dark solitons within the context of their existence as solutions of differential equations. Of course since difference equations are a closer form of reality to the dynamics of physical systems, these excitations must find some counterpart in the solutions of difference equations, and this is the case to a large extent. It should also be noted that some of the nonlinear pulse and dip solutions found in nonlinear difference equations are not straightforwardly represented by difference equations. Consider for example pulses and dips with a rapidly varying phase between neighboring sites. A special treatment may be required to handle the phase of these types of excitations. In this section, it will be shown how the general difference equation form given in (7.8) reduces to a nonlinear differential equation in the continuum limit. The reader is then referred to chapters 5 and 6 for the theory of the treatment of the resulting nonlinear differential equations [11].

The discrete chain soliton solutions discussed in this chapter are useful in the study of modes which are highly localized pulses or dips forming bright, dark, and gray solitons. In particular, our treatments of the discrete system have focused on narrow pulses and dips which are rapidly varying over the chain lattice. These highly localized forms are one limit of the solitons that are of interest in these nonlinear systems. Bright, dark, and gray solitons can also occur as broad peaks and dips, and we now turn to a discussion of these types of excitations. The theory of broad excitations is best handled in a differential equation format obtained in the continuum limit of the nonlinear systems.

In the continuum the discrete variables s_n of the difference equations are treated as continuous variables $s(x)$ where $s_n = s(x)\,|_{x=na}$. If the wave function of the excitations in the difference equation is slowly varying from site to nearest neighbor site along the chain, then in terms of the lattice constant of the one-dimensional lattice, a, a Taylor expansion gives

$$s_{n\pm1} = \left[s \pm a\frac{\partial s}{\partial x} + \frac{1}{2}a^2\frac{\partial^2 s}{\partial x^2} + \cdots \right]_{x=na}. \tag{7.32}$$

Substituting these forms into (7.8) generates the differential equation

$$B(\omega)a^2\frac{d^2 s}{dx^2} + [A(\omega) + 2B(\omega)]s + [C(\omega) + 2D(\omega)]\,|\,s\,|^2\,s = 0. \tag{7.33}$$

Here only the second order derivatives are retained, the coefficients $C(\omega)$ and $D(\omega)$ are small parameters, and $s(x)$ is slowly varying in x. For discussions of the soliton solutions found in various continuum limit equations obtained from the nonlinear generalization of (7.1) through (7.4) see chapters 5 and 6.

References

[1] Giri P, Choudhary K, Sen Gupta A, Bandyopadhyay A K and McGurn A R 2011 Klein–Gordon equation approach to nonlinear split-ring resonator based metamaterials: one-dimensional systems *Phys. Rev.* B **84** 155429

[2] Eleftheriou M, Lazarides N and Tsironis G P 2008 Magnetoinductive breathers in metamaterials *Phys. Rev.* E **77** 036608

[3] Eleftheriades G V 2009 EM transmission-line metamaterials *Mater. Today* **12** 30

[4] Kourakis I, Lazarides N and Tsironis G P 2007 Self-focusing and envelope pulse generation in nonlinear magnetic metamaterials *Phys. Rev.* E **75** 067601

[5] Lapine M, Gorkunov M and Ringhofer K H 2003 Nonlinearity of a metamaterial arising from diode inserts into resonant conductive elements *Phys. Rev.* E **67** 065601

[6] Shadrivov I V, Morrison S K and Kivshar Y 2006 Tunable split-ring resonators for nonlinear negative-index metamaterials *Opt. Express* **14** 9344

[7] Tamayama Y, Nakanishi T and Kitano M 2013 Observation of modulation instability in a nonlinear magnetoinductive waveguide *Phys. Rev.* B **87** 195123
[8] McGurn A R 1996 Green's-function theory for row and periodic defect arrays in photonic band structures *Phys. Rev.* B **53** 7059
[9] Dauxois T and Peyrard M 2006 *Physics of Solitons* (Cambridge: Cambridge University Press)
[10] Sievers A J and Page J B 1995 *Dynamical Properties of Solids* ed G K Horton and A A Maradudin (Amsterdam: North-Holland)
[11] McGurn A R 2013 Barrier transmission map of one-dimensional nonlinear split-ring-resonator-based metamaterials: bright, dark, and gray soliton resonances *Phys. Rev.* B **88** 155110
[12] McGurn A R 2008 Transmission through nonlinear barriers *Phys. Rev.* B **77** 115105
[13] Marder M P 2010 *Condensed Matter Physics* 2nd edn (Hoboken, NJ: Wiley)
[14] McGurn A R 1999 Intrinsic localized modes in nonlinear photonic crystal waveguides *Phys. Lett.* A **251** 322
[15] McGurn A R 2003 Intrinsic localized modes in photonic crystal circuits *Chaos* **13** 754

Arthur R McGurn

Chapter 8

Additional aspects and applications of nonlinearity

Some important aspects of nonlinearity in nano-photonics are related to optical bistability and to the waveguide transmission characteristics through barriers of nonlinear optical media. Optical bistability is a well-known phenomenon in which the Kerr nonlinearity of a system allows for the existence of multiple states of an optical system to simultaneously exist as part of the system solution set. These multiple solutions cause the systems to exhibit hysteresis effects similar to those observed in permanent magnetic systems. Specifically, in optical bistability the final state of a Kerr nonlinear system depends on the past history of the system as well as the initial state of the system. The possibility of multiple solutions is very important as a basic phenomenon and finds a use in the design of optical switches and transistors.

Another interesting phenomenon is the anomalies found in the transmission of guided modes through barriers of nonlinear optical media inserted within the waveguide channel, i.e., the properties of nonlinear waveguide barriers. Waveguide barriers composed of linear optical media exhibit well known Fabry–Perot resonant transmission anomalies which are related to the plane wave modes of the linear barrier media [1]. These are well understood from the theory of lasers where they form the basis of lasing cavities. In the case of waveguide barriers of Kerr nonlinear media, additional anomalies are found to come from the resonant excitation of various types of soliton modes within the barrier media. These are in addition to the transmission anomalies which arise from the renormalized plane wave modes of the nonlinear barrier. Both types of resonant anomalies must be taken into consideration with the introduction into optical circuits of devices composed of nonlinear media.

In the following, the basics of these two aspects of the possible use of nonlinearity for future device designs will be outline with examples. This will be carried out using

doi:10.1088/978-0-7503-3579-9ch8

the simple difference equation models for photonic crystal and split ring resonator (SRR) metamaterial waveguide systems.

8.1 Optical bistability

A simple system which illustrates the principles of optical bistability [2–4] is a waveguide composed of linear optical media which contains a single site composed of Kerr nonlinear optical media, i.e., a Kerr nonlinear impurity site [5]. The nonlinear site replaces one of the linear media sites of the waveguide which otherwise consists solely of linear media. For this type of system the problem of the scattering of a waveguide mode from the nonlinear site is studied. As a result of the interaction of the incident guided mode at the nonlinear impurity site, reflected and transmitted modes are generated at the impurity, and these three modes are related by the laws of electrodynamics. Optical bistability occurs in the system when a single transmitted mode solution is found to correspond to multiple different solutions for the incident and reflected waves. Bistability only arises from the presence of nonlinearity in the system, originating in the nonlinear nature of the dynamical equations of the system. Nonlinear equations, unlike linear equations, allow for the possibility of multiple solutions.

The waveguide scattering problem for a single Kerr site impurity will be illustrated by a treatment for photonic crystal waveguides within the context of the difference equation formulation. Within this formulation the origins of the optical bistability are readily seen. For the photonic crystal model in (7.3) and (7.18) the difference equations of the proposed system are described by [5]

$$E_n = g_l[\alpha E_n + \beta(E_{n-1} + E_{n+1})], \tag{8.1a}$$

for $n \leqslant -2$ or $n \geqslant 2$

$$E_{\pm 1} = g_l[\alpha E_{\pm 1} + \beta E_{\pm 2}] + g\beta[E_0 + \lambda \mid E_0 \mid^2 E_0], \tag{8.1b}$$

and

$$E_0 = g\alpha[E_0 + \lambda \mid E_0 \mid^2 E_0] + g_l\beta(E_1 + E_{-1}). \tag{8.1c}$$

Equation (8.1a) is the difference equation for the linear medium waveguide channel surrounding the Kerr impurity site which is located at $n = 0$, while (8.1b) and (8.1c) describe the Kerr impurity site and its coupling to its neighboring linear medium sites. The object will be to study the one-dimensional scattering problem of an incident guided mode on the left of the impurity and a transmitted wave on the right of the Kerr impurity.

The scattering solutions of (8.1) for a guided mode incident on the Kerr nonlinear impurity from its left are of the form

$$E_n = u\mathrm{e}^{\mathrm{i}kn} + r\mathrm{e}^{-\mathrm{i}kn}, \tag{8.2a}$$

for $n \leqslant -1$

$$E_n = t\mathrm{e}^{\mathrm{i}kn}, \tag{8.2b}$$

for $n \geqslant 1$ and at $n = 0$

$$E_0 = s. \tag{8.2c}$$

Equation (8.2a) is a linear combination of the incident and scattered waves, (8.2b) is the transmitted wave, and (8.2c) is the wave at the nonlinear site. The system is studied by substituting (8.2) into (8.1) and solving for u, r, s in terms of t.

Substituting (8.2) into (8.1), the transmission coefficient of the guided mode scattering is given by [5]

$$T = \left| \frac{t}{u} \right|^2 = \frac{4\sin^2 k}{\left| \left(\frac{\alpha}{\beta} + 2\cos k \right)\left(\frac{s}{t} - 1 \right) - 2\mathrm{i} \sin k \right|^2}, \tag{8.3a}$$

where

$$\lambda \mid s \mid^2 s + s - \frac{g_l}{g} t = 0, \tag{8.3b}$$

and

$$1 = g_l[\alpha + 2\beta \cos k]. \tag{8.3c}$$

The reflection coefficient is $R = \left| \frac{r}{u} \right|^2$ and the transmission and reflection coefficients satisfy $T + R = 1$. From (8.3a) it is seen that, for fixed t and wave number k, the transmission through the site depends on the solutions for s from (8.3b). This is the key to the optical bistability properties of the system. The optical bistability arises from the possibility of multiple solutions of the cubic equation in (8.3b) so that for a fixed t there are multiple $\frac{t}{u}$ and T.

Specifically, the coefficients in (8.3b) are all real so that s must be a real solution of

$$s^3 + \frac{1}{\lambda} s - \frac{g_l}{\lambda g} t = 0. \tag{8.4}$$

From basic algebra, a cubic equation of the form

$$x^3 + a_2 s + a_3 = 0. \tag{8.5a}$$

has a discriminant

$$D = \left(\frac{a_2}{3} \right)^3 + \left(\frac{a_3}{2} \right)^2, \tag{8.5b}$$

which determines the multiplicity of its solutions. For $D > 0$ (8.5a) has a single real solution for x, for $D < 0$ it has three real solutions for x, and for $D = 0$ it has three real solutions for x, two of which are equal. The bistability of the transmission coefficients in (8.3) follow from applying these considerations of the discriminant to (8.3) through (8.5).

In figure 8.1 the optical transmission coefficient is computed for an example of the waveguide with a Kerr nonlinear impurity and is plotted versus the Kerr intensity

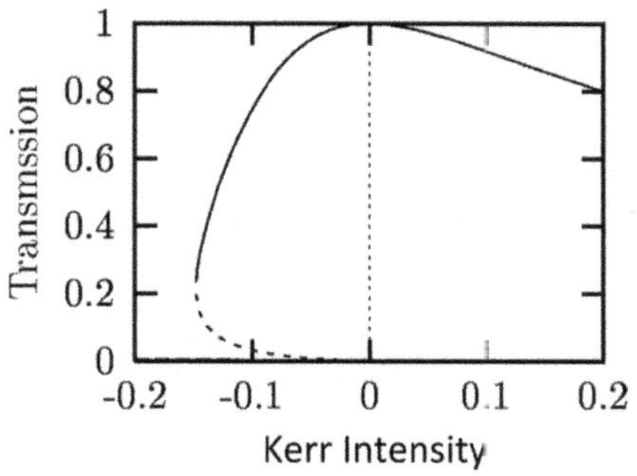

Figure 8.1. Optical bistability plot of transmission versus the Kerr intensity for a single frequency mode. Part of the solution skims the negative x-axis and may be difficult to resolve on the scale of the plot. Reproduced with permission from [6]. Copyright 2003 American Institute of Physics.

defined as $\lambda \mid t \mid^2$. It gives a typical example of optical bistability, showing the effects of changing the intensity of the transmitted wave on the transmission coefficient [5]. The details of the fixed parameters α, β, g, g_l, and k of the particular system can be found in the literature where the figure originally appears. Here only the general features of the plot, which are common to all optically bistable systems, are discussed.

For the Kerr intensity equal to zero (i.e., $\lambda = 0$) the system is a linear medium system. The parameters of the particular system plotted have been arranged so that in the linear limit the transmission coefficient through the nonlinear site is one and there is no reflected wave [5, 6]. As the Kerr intensity becomes increasingly positive the transmission coefficient decreases as the nonlinearity increases the dielectric contrast between the impurity site and that of the linear medium waveguide channel. This creates a reflected wave of increasing amplitude. The transmission coefficient follows the solid line in the plot, moving towards the right in the figure. Likewise, upon making the Kerr intensity increasingly more negative the transmission coefficient decreases for the same reason, following the solid line in the plot towards the left-hand side of the plot. Near a Kerr intensity of −0.15, however, the solid line ends and with increasing Kerr intensity the transmission coefficient falls off the curve. With further increasing Kerr intensity it resumes motion to the left of the figure on a line of solutions which skims the x-axis, being just above the x-axis. (Unfortunately, for this plot the skinning line of solutions is hard to resolve from the x-axis on the plot.) Upon reversing the trajectory, with decreasingly negative Kerr intensity the solution follows the line skimming the x-axis until it reaches the zero of Kerr intensity. Here the solution jumps to the top curve which passes through the linear limit solution of the system.

The system is seen to exhibit a hysteresis effect similar to that found in permanent magnetic systems. The origin of the region of multiple solutions is the multiple solutions of (8.4) which are predicted from the cubic discriminant in (8.5b). It should be noted that the short dashed line in figure 8.1 is also part of the solution set of the transmission coefficient. This branch, however, does not naturally connect to the transmission coefficient branches at large positive or large negative Kerr intensities. Rather, the two branches of solution containing these limits are disconnected, with the short dashed solution acting as an interpolation between the two branches. The short dashed branch interpolation represents a set of unstable solutions that need not be considered for most purposes [5, 6].

The optical bistability property exhibited in figure 8.1 is interesting from a technological standpoint as the dependence of the Kerr dielectric constant on the intensity of the field applied to it allows a single frequency signal to modulate itself through amplitude changes in its field. That is to say that by varying the Kerr intensity of the signal, its transmission and reflection properties are changed. Such types of modulations can be used as a design basis for various switching devices for electro-optical applications.

Another aspect of the Kerr dependence of the dielectric constant on the intensities of the fields applied to it is that the Kerr interaction can be used to have waves of different frequencies interact with one another [5]. The results in figure 8.2 illustrate this point. Here the linear medium waveguide with a single Kerr medium site similar

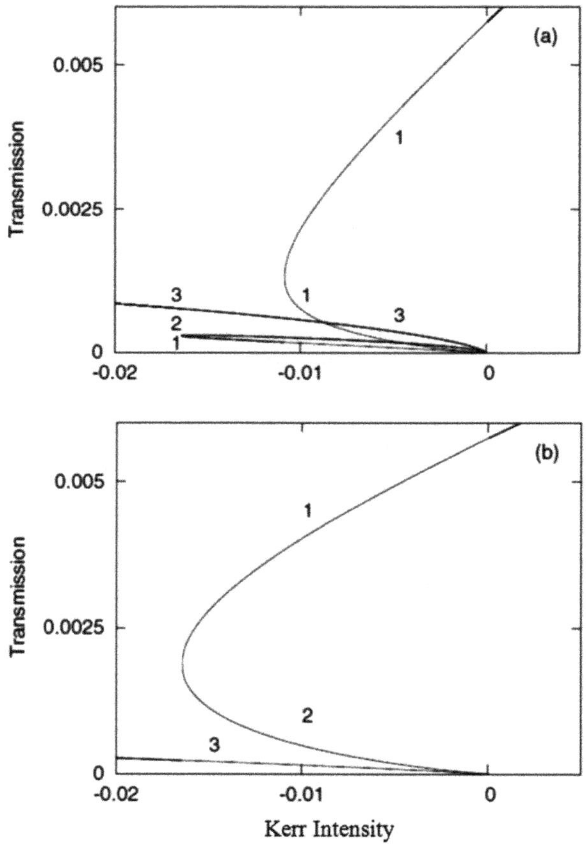

Figure 8.2. Optical multi-stability plots of transmission versus Kerr intensity for two guided modes: (a) results for the mode with a self-interaction as well as an interaction with the other mode in the system, and (b) results for the mode with only a self-interaction. Reproduced with permission from [5]. Copyright 2015 American Physical Society.

to that considered in (8.1), (8.3) and figure 8.1 is treated. Now, however, two modes of closely spaced frequencies are incident from the left on the impurity. Both are in part transmitted and in part reflected at the impurity site. To facilitate the study one incident mode was taken to only interact with itself through the Kerr media. This is the 'self-only' interacting mode. The second incident mode, however, was taken to interact, through the Kerr media, equally with itself and with the other mode in the system. This is then the 'total' interacting mode. The Kerr interaction was taken to be the same for both modes and the transmission coefficient of each mode is studied as a function of the Kerr interaction.

Again the reader is referred to the literature for the details of the difference equation parameters, the dielectric constants, and the wave number of the modes [5]. These parameters are all fixed so that the only variation is in the Kerr intensity. Here only the qualitative behaviors of the system are of interest.

Figure 8.2 presents the transmission coefficients of the two modes versus the Kerr intensity which is the same for each mode. Figure 8.2(a) is the transmission coefficient of the 'total' interacting mode and figure 8.2(b) is the transmission coefficient of the 'self-only' interacting mode. The curve in figure 8.2(b) is, as it should be, a version of the bistable system presented in figure 8.1. This is due to the fact that the mode in figure 8.2(b) does not see the other mode in the system. The branches of the solutions in figure 8.2 are labeled 1, 2, and 3, where curves in the two plots with the same labels are paired as solution sets of the equations of motion for the two modes. These equations are nonlinear so they support the possibilities of multiple pairings of the solutions.

In figure 8.2 the system is linear when the Kerr intensity is zero and both frequencies have a transmission coefficient of approximately 0.006. In the present system the linear impurity is not matched to the linear dielectric constant of the sites of the waveguide channel, and this accounts for the presence of a reflected wave for both frequencies in the linear limit of the system. Increasing the Kerr intensity either in a positive or negative sense, both modes move continuously from the linear system on the curves labeled 1. In figure 8.2(a) moving to the left, the solution travels along 1 and eventually, near Kerr intensity = −0.01, falls off the curve labeled 1 and lands on the second lower curve labeled 1. Eventually near Kerr intensity = −0.016 the solution jumps up to the curve labeled 3. During this process the solution in figure 8.2(b) moves to the left from the linear limit on the curve labeled 1. It follows the curve, falling off near Kerr intensity = −0.016 to land on the curve labeled 3. During these processes the labels of the solutions in both figures must agree. Reversing this process, returning to the right, the solutions in both plots follow the curves labeled 3 and near the Kerr intensity = 0 jump to the linear limit on curve 1. The presence of two interacting modes in the system is seen to allow them to modify each other's motions.

Both of the systems presented here exhibit multiple solutions at negative Kerr intensities, but there are also systems that exhibit multiple solutions at positive Kerr intensities [5]. All of these types of systems are readily understood within the context of the difference equation formulation. This is done by studying the discriminant of cubic equations as given in (8.5) because all of the various difference equation formulations exhibit multiple solutions based on the solutions of cubic polynomials.

It should be noted that similar multiple solution properties to those discussed above have been worked out for the SRR metamaterial waveguides. Considerations of both photonic crystal and SRR metamaterial waveguides with both in-channel and off-channel Kerr nonlinear medium impurities have also been made, and many of these solutions exhibit bistability and multiple solution properties [5]. A comparison of the bistability and multiple solution properties of the photonic crystal and SRR metamaterial problems indicates, due to differences in the structure of the difference equations describing these two materials, that the two types of systems display qualitatively different solution sets. The reader is referred to the literature for the complete treatment of photonic crystal and SRR metamaterial waveguide problems.

8.2 Transmission through barriers of nonlinear optical media

A related problem to those of nonlinear site impurities in waveguides are systems with a number of nonlinear sites introduced into an otherwise linear medium waveguide [7, 8]. These are barriers of nonlinear optical media, and the focus of these problems will be on single frequency waveguide modes. Such barriers introduce new behaviors into the system arising through the introduction of Fabry–Perot plane wave resonances and through the resonant excitation of soliton modes in the barrier media [1]. In addition, soliton modes allow for the creation of intense electromagnetic fields localized within the nonlinear barrier sites. The existences of all of these types of resonant modes must be understood to account for the effects of the introduction of devices formed from nonlinear optical media within opto-electronic circuits. In the following some of these effects will be briefly illustrated for SRR metamaterial waveguides. The reader is referred to the literature for a detailed treatment of nonlinear barrier systems in both photonic crystals and SRR metamaterials. Here only a basic outline is presented.

For the illustration, a problem similar to the single site scattering problem in section 8.1 is treated for an SRR metamaterial waveguide [7]. A linear medium SRR metamaterial waveguide characterized by the inductive nearest neighbor couplings, λ, carries a single frequency incident guided mode of wave number, k. The incident guided mode arrives from the left of the barrier and is scattered by the nonlinear media of the barrier. The SRRs of the barrier are coupled by nearest neighbor couplings λ', and the Kerr nonlinear barrier medium is characterized by the interaction parameter $\frac{\alpha}{\varepsilon_l}$. The details of the numerical parameters used in the calculation and of the difference equations of the system are not given here. The reader is referred to the literature for the details of these [7]. Here only the qualitative natures of the solutions, which are common to many types of system described by difference equations, are discussed.

The scattering solutions from the barrier are obtained from the difference equations by applying scattering solution boundary conditions. To the left of the barrier the scattering solutions are composed of an incident and reflected wave of the form [7]

$$q_n = u\mathrm{e}^{\mathrm{i}kn} + r\mathrm{e}^{-\mathrm{i}kn}, \tag{8.6a}$$

and to the right of the barrier the scattering solutions are of the form of a transmitted wave

$$q_n = te^{ikn}. \tag{8.6b}$$

As in the case of the optical bistability discussions for a single frequency incident mode, a value of t for the transmitted wave is fixed and (u, r) are solved for in terms of t. From these solutions, the transmission and reflection coefficients, $T = \left| \frac{t}{u} \right|^2$ and $R = \left| \frac{r}{u} \right|^2$, are computed, respectively, where for the conservative system it is found that $T + R = 1$. In the following, it will be shown that for the solutions obtained various transmission resonances in the barrier scattering result. These resonances arise from the excitation of Fabry–Perot and soliton modes within the barrier media which facilitate the passage of energy through the barrier. This has a significant effect on the propagation characteristics of barrier devices when placed in photonic circuits as the resonant excitations also affect the flow of electromagnetic energy in the entire circuit.

In figure 8.3 plots of the location of the transmission maxima within the two-dimensional space of the nonlinear barrier parameters $(\frac{\alpha}{\varepsilon_l} \mid t \mid^2, \lambda')$ is presented for a barrier of nine Kerr nonlinear medium sites located within a waveguide of linear media [7]. The barrier is formed by replacing nine consecutive sites within the linear medium waveguide, and the two parameters $(\frac{\alpha}{\varepsilon_l} \mid t \mid^2, \lambda')$ characterize, respectively, the intensity of the nonlinearity of the barrier material and the inductive couplings between the nearest neighboring SRR sites of the nonlinear medium barrier. In the linear medium waveguide the parameters λ and k of the inductive couplings and the wave number of the guided modes, respectively, are fixed over the entire plot in figure 8.3 so that $\left(\frac{\alpha}{\varepsilon_l} \mid t \mid^2, \lambda'\right)$ are the only parameters that are allowed to vary in determining the plot.

The lines of resonant maxima in figure 8.3 are determined as follows. For each value of $\frac{\alpha}{\varepsilon_l} \mid t \mid^2$, the transmission coefficient of the system is computed as a function of λ' and the positions of the resulting transmission coefficient maxima are

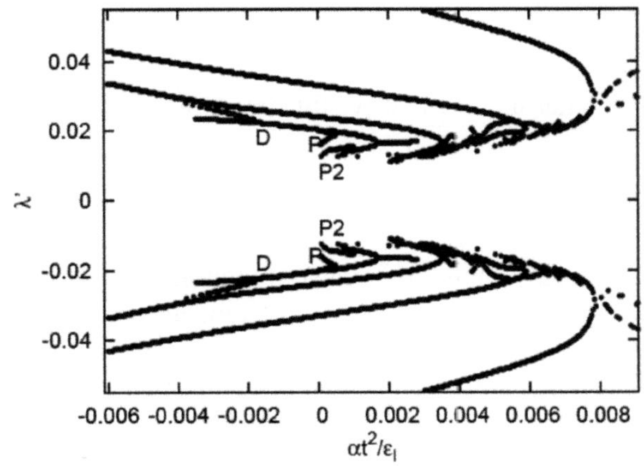

Figure 8.3. Plot of the positions of the resonant transmission peaks in the $(\lambda', \alpha t^2/\varepsilon_l)$ plane. Reproduced with permission from [7]. Copyright 2013 American Physical Society.

determined. The resulting positions of the maxima in the space of $\left(\frac{\alpha}{\varepsilon_l} \mid t \mid^2, \lambda'\right)$ are then presented as points on the plot. The maxima can be shown to arise from the resonant excitations of the natural modes of the nonlinear barrier media. In the case of a barrier of linear dielectric media (i.e., for $\frac{\alpha}{\varepsilon_l} \mid t \mid^2 = 0$) the resonantly excited modes are the Fabry–Perot modes, familiar from the study of laser cavities. For barriers of nonlinear media, in addition to supporting Fabry–Perot modes that are renormalized by the nonlinearity, the barrier also supports the excitation of soliton resonant modes. These only exist in the presence of nonlinearity. In both systems the resonantly excited modes of the barrier facilitate the passage of electromagnetic energy through the barrier media, showing up as transmission maxima [7].

In figure 8.3, $\frac{\alpha}{\varepsilon_l} \mid t \mid^2 = 0$ marks the limit of the system in which the barrier media is a linear medium and the resonant excitations at $\left(\frac{\alpha}{\varepsilon_l} \mid t \mid^2, \lambda'\right) = (0, \lambda')$ are the Fabry–Perot resonance of the linear limit. Upon going to either the right or left of $\frac{\alpha}{\varepsilon_l} \mid t \mid^2 = 0$ the barrier medium system becomes progressively nonlinear. As the nonlinearity increases, the Fabry–Perot resonances are renormalized with the increasing nonlinear. During the renormalization process the wave functions of the Fabry–Perot modes continue to retain their distinctive morphologies as found in the linear system. In addition, with increasing nonlinearity new lines of excitations appear in the diagram. These represent new types of barrier excitations which can be identified with soliton modes resonantly excited within the nonlinear barrier media. We turn to focus the discussion on the nature of the soliton barrier resonances.

In figure 8.3 bright soliton modes are found to be associated with two branches of modes in the figure that are labeled with the letter P [7]. The P branches of modes are absent from the system at the $\frac{\alpha}{\varepsilon_l} \mid t \mid^2 = 0$ linear limit of the system. Both P branches begin at a very small positive value of $\frac{\alpha}{\varepsilon_l} \mid t \mid^2$ and the branches continue in the system with increasing positive $\frac{\alpha}{\varepsilon_l} \mid t \mid^2$ until they join up with branches of renormalized Fabry–Perot modes. The identification of these modes with bright solitons can be made by comparing their wave functions and frequencies with those of the bright soliton solutions for an infinite chain of Kerr medium sites with the parameterization of the sites in the barrier. This is now addressed.

In figure 8.4(a) the intensity of the wave functions at a small fixed $\frac{\alpha}{\varepsilon_l} \mid t \mid^2$ are plotted versus the nonlinear barrier sites, labeled $l = 1, 2, \ldots, 9$. The value of $\frac{\alpha}{\varepsilon_l} \mid t \mid^2$ is small positive and at the left-hand end of the P branches, and the wave functions shown are computed at the discrete values of λ' at which their resonant transmission maxima are observed. In the intensity plots, the wave function intensities are found to be independent of the sign of λ' but differ for different absolute magnitudes of λ'. The lower intensity modes in the plot are the renormalized Fabry–Perot modes, and the single mode with the large peak is the resonantly excited bright soliton mode. The intensity of the field at the soliton peak is seen to be much larger than that at the edges of the pulse which are comparable to the fields in the linear waveguide. The field generated at the soliton peak is greatly enhanced from the fields of the applied waveguide mode.

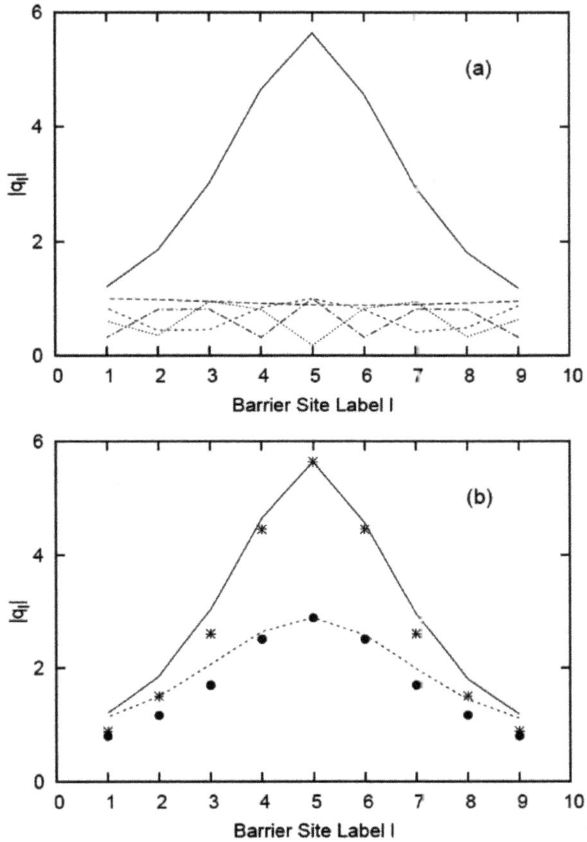

Figure 8.4. Plot of the absolute value of mode amplitudes versus barrier site label: (a) the large soliton peak compared to the amplitudes of the renormalized plane waves, and (b) comparison of the resonantly excited barrier soliton modes (connected by lines) and results from soliton modes in the corresponding infinite chain (points). Reproduced with permission from [7]. Copyright 2013 American Physical Society.

In figure 8.4(b) a comparison is made between the intensities of the resonantly excited fields in the nine site barrier with the intensities of the fields of the corresponding bright soliton solutions in the infinite Kerr nonlinear waveguide [7]. The excited barrier modes are indicated by the connecting lines in the plot, while the solutions for the infinite Kerr nonlinear waveguide are shown by the plotted points. The comparison is made at two different $\frac{\alpha}{\varepsilon_l} \mid t \mid^2$ along the P labeled line. The good agreement between the two solutions confirms the nature of the modes resonantly excited within the nonlinear barrier media as solitons.

In addition to the P branches two branches have been labeled P2 and another set of branches have been labeled D. The P2 branches are identified with the excitation of two (i.e., multiple) bright soliton pulses within the barrier, and the branches labeled D are identified with the excitation of dark solitons within the barrier. To confirm this identification in the case of dark solitons, in figure 8.5 the wave function intensities on and around the branch of modes labeled D in figure 8.3 are shown [7].

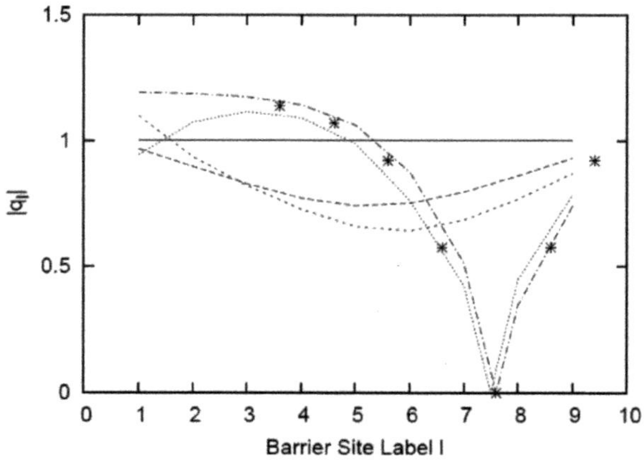

Figure 8.5. Dark soliton wave function versus barrier label. The lines are for the resonantly excited barrier modes and the points are the corresponding solutions for the dark solitons in the infinite chain. Reproduced with permission from [7]. Copyright 2013 American Physical Society.

(As with the bright solitons, the wave function intensities are the same for systems with the same $|\lambda'|$.) The D branch is seen as a line of modes in figure 8.3 which begins on a line of Fabry–Perot modes at negative $\frac{\alpha}{\varepsilon_l} \mid t \mid^2$ and ends on another line of Fabry–Perot modes at negative $\frac{\alpha}{\varepsilon_l} \mid t \mid^2$. It is a connecting line between the two Fabry–Perot branches. In figure 8.5 the dark soliton on the connecting D branch is seen as a line with an intensity dip. The line with the intensity dip is the resonant mode intensity in the nonlinear barrier. On this line has been plotted points indicating the corresponding dark soliton solution in the infinite nonlinear waveguide. The correspondence of the two solutions is good and confirms the dark soliton nature of the resonantly excited barrier mode. The other lines in figure 8.5 are for modes on the neighboring lines of renormalized Fabry–Perot modes, and the reader is referred to the literature for a discussion of these.

References

[1] Smith W J 2007 *Modern Optical Engineering* 4th edn (New York: McGraw-Hill)

[2] Boyd R W 2003 *Nonlinear Optics* 2nd edn (Amsterdam: Academic)

[3] Mills D L 1998 *Nonlinear Optics* (Berlin: Springer)

[4] Banerjee P P 2004 *Nonlinear Optics* (New York: Dekker)

[5] Rai B and McGurn A R 2015 Multiplicity of transmission coefficients in photonic crystal and split ring resonator waveguides with Kerr nonlinear impurities *Phys. Rev.* B **91** 085113

[6] McGurn A R 2003 Intrinsic localized modes in photonic crystal circuits *Chaos* **13** 754

[7] McGurn A R 2013 Barrier transmission map of one-dimensional nonlinear split-ring-resonator-based metamaterials: bright, dark, and gray soliton resonances *Phys. Rev.* B **88** 155110

[8] McGurn A R 2008 Transmission through nonlinear barriers *Phys. Rev.* B **77** 115105

Arthur R McGurn

Chapter 9

Basic applications: Kerr media, second harmonic generation, parametric oscillation and amplification

In this chapter a discussion is given of the basic phenomena and problems arising in applications of nonlinear optics in opto-electronic systems for the frequency conversion of light. Examples are the generation of a second harmonic from the application to a nonlinear medium of light at a fundamental frequency and the interactions of multiple different frequencies with nonlinear materials so as to create various sum and difference combinations of light in the system. The idea is to focus only on the fundamental physical properties and types of computations that have been performed, with a discussion given of their role in understanding the design requirements of devices. In this regard, the present treatment compliments some of the earlier discussions of methodology given in chapters 3 and 4, which primarily considered linear media and later chapters in which the focus was on Kerr media. A comprehensive review of nonlinear applications and their associated physical phenomena, however, is not given in favor of a pedagogical presentation. The interested reader is referred to the literature for further, more detailed, reviews. In addition, some addition discussions of Kerr nonlinearity topics are presented as preliminary to chapter 10 materials.

A focus in electro-optics is on numerical simulations involving finite-difference time-domain methods [1]. These are responsible for considerable theoretical work in the opto-electronics of higher-dimensional systems. Such simulations are necessary from the standpoint of engineering where a great variety of systems need to be accurately and successfully considered through the application of standardized simulation techniques [1]. Often, however, the detailed results of numerical simulations do not give simple, easily understandable, explanations of the physics of the nonlinear

doi:10.1088/978-0-7503-3579-9ch9

systems studied. This is why the earlier discussions given in this book have emphasized analytical models and the present chapter will continue this focus.

An initial treatment of this chapter is a continuation of some discussions of linear and Kerr nonlinear media in a presentation on transfer matrix methods. This technique is often used to study one-dimensional (layered) media and waveguides [2] and usually involves exact solutions of one-dimensional systems. Such one-dimensional treatments in many instances exhibit the qualitative characteristics found in higher-dimensional systems but allow for exact solutions of the system dynamics. When applicable to the solution of optical systems, transfer matrix methods are easier to use than finite-difference time-domain methods, as well as being faster and more accurate methods. Transfer matrix methods are applicable to a wide range of devices that are of importance as circuit components in opto-electric engineering and have a rich literature [2]. The transfer matrix techniques will be considered again in chapter 10 in the context of the study of disorder and Anderson localization where they will be illustrated in detail for specific models involving linear and nonlinear media.

Aside from the continued discussion of interesting aspects of Kerr nonlinearity, the bulk of the present chapter will present a new focus on problems involving the generation of higher or lower harmonics of radiation. In these problems, radiation of a frequency ω can generate or be generated by the interaction of light of different frequencies ω_1 and ω_2 in a nonlinear medium. This interaction is facilitated by the basic physical properties of the nonlinear optical medium. The problem of the generation of second harmonics is an example of such a process, and the discussion given here will expand these considerations to treat further frequency combinations of light.

In the following, the generation of second harmonics of light is treated in detail. The discussions focus on the limitations placed on the physical effect arising from the crystal symmetry and the macroscopic geometry of samples of nonlinear media. These limitations include discussions of phase matching effects and treatments related to the crystal symmetry placed on materials that generate second harmonics. Phase match effects are concerned with the interference of second harmonics of radiation generated at different points in a macroscopic medium. Crystal symmetry enters at a microscopic level into the generation of second harmonics as second harmonics can only be generated in the absence of certain crystalline group symmetries. Engineered material are design to relieve some of these limitations on the generation of second harmonics. In this regard, the applications of photonic crystals and metamaterials to enhance second harmonic generation is discussed with references to the current literature.

Related to the problem of second harmonics is that of the generation of other harmonic frequencies of light in nonlinear media. In these processes the nonlinear medium causes interactions involving three or more different frequency components. The problems involved with these interactions are associated with important types of processes known as parametric oscillation and parametric amplification. These are found as useful effects in photonic crystals and metamaterials and in their applications in optical technology. A number of such technological applications which are intimately related to these phenomena are briefly reviewed. The focus in this chapter, however, is on a pedagogical introduction to these phenomena and the basics of their applications. For a detailed review, the reader is again referred to the literature.

In association with problems involving Kerr nonlinearity in photonic crystals and metamaterials some general discussions are presented of nonlinear fiber optic waveguides. These include the solution of soliton modes in Kerr nonlinear fiber waveguides, and a basic treatment of these modes is presented here. In addition, some discussion of parametric oscillation and amplification in optical fibers is also given. The discussions of solitons in fiber optics, as do those of parametric oscillation and amplification, historically preceded their treatment in photonic crystals and metamaterials. These topics in fiber optics form another important component in the study of device technologies and are included here as an introduction to the reader and as a comparison with their counterparts in photonic crystal and metamaterial systems.

In regard to our earlier studies of metamaterials some additional discussion is given here of a number of developments made involving the perfection of split ring resonator technology. These include engineering designs introduced to increase their performance in various frequency regions of technological interest.

For the following presentation, first the basics of transfer matrix methods [2] will be outlined and some of the results from these studies described. After this a brief discussion is given of finite-difference time-domain [1] treatments applied to a select set of examples in the nonlinear optics of nano-photonic systems. Second harmonic generation is then discussed followed by treatments of parametric oscillation and amplification. The chapter concludes with an introduction to nonlinear optical fibers and a brief summary of recent developments in the technology of split ring resonators.

9.1 Transfer matrix methods: linear and Kerr nonlinear media

Transfer matrix methods for systems involving linear and Kerr nonlinear media are generally applicable to problems involving the study of layered systems. These are often composed of dielectric slabs which are stacked so that the total system is translational invariant in one plane in space but not along the axis perpendicular to the plane. The translation symmetry in the plane simplifies the properties of the system. It allows the problem of finding the modal wave functions to reduce to a one-dimensional problem directed along the axis perpendicular to the symmetry plane. The reduced problem is written in terms of a product of matrices (i.e., the transfer matrices of the system) which generates the wave function by projecting it along the length of the chain.

To understand the method of transfer matrix, begin by considering a slab or single layer of dielectric medium for electromagnetic modes propagating in the direction normal to the surfaces of the slab [2]. The solutions of the more general layered media can then be expressed in terms of these single slab solutions. Take the axis normal to the slab surfaces to be the x-axis and the surfaces of the slab to be located at $x = 0$ and $x = d$ (see figure 9.1 for a schematic of the slab surrounded by vacuum). Note that for the consideration to follow, the slab is taken to be surrounded by a single linear dielectric medium.

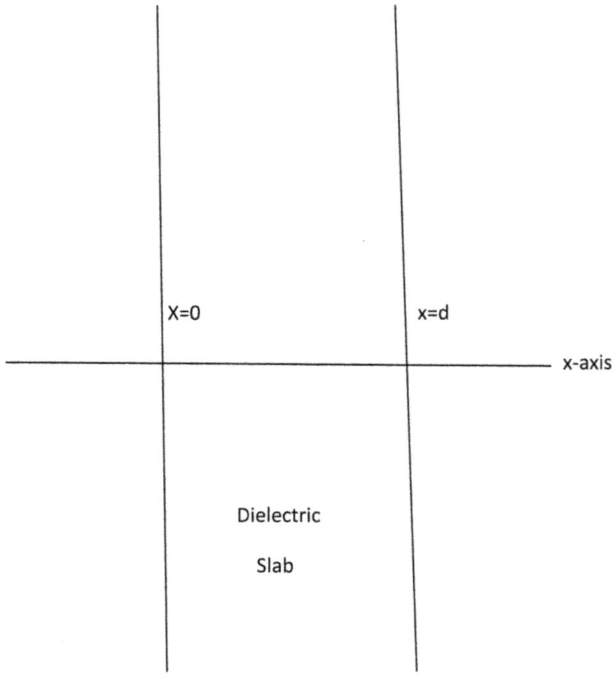

Figure 9.1. Single dielectric slab with surfaces at $x = 0$ and $x = d$.

For waves moving along the x-axis, to the left and right of the slab the solutions must be composed of a linear combination of both left and right propagating plane waves. Consequently, to the left of the slab for $x < 0$ the electric fields are of the form [2]

$$E_L = A_1 e^{ikx} + A_2 e^{-ikx}, \tag{9.1a}$$

and to the right of the slab for $d < x$ the electric fields are of the form

$$E_R = B_1 e^{ikx} + B_2 e^{-ikx}. \tag{9.1b}$$

By matching these solutions with the solutions within the slab (this is done by applying the electromagnetic boundary conditions at the surfaces of the slab) the A_i and B_i coefficients for $i = 1, 2$ are found to be related to one another by a matrix equation of the form [2]

$$\begin{vmatrix} M_{11} & M_{12} \\ M_{21} & M_{22} \end{vmatrix} \begin{vmatrix} B_1 \\ B_2 \end{vmatrix} = \begin{vmatrix} A_1 \\ A_2 \end{vmatrix}. \tag{9.2}$$

Here $M_{i,j}$ for $i, j = 1,2$ are written in terms of the dielectric parameters and the positions of the two surfaces of the slab [2–6]. In the case of a slab of linear dielectric medium the $M_{i,j}$ are expressed exactly in terms of algebras involving plane wave forms, but in the case of a slab of nonlinear dielectric medium the $M_{i,j}$ are expressed exactly in terms of algebras involving elliptic integrals [5–8]. Given any two of the set of A_i and B_i coefficients for $i = 1, 2$ the remaining two members of the set can be found. In this way, the result in (9.2) can be used to write the modal solutions for a general layering of slabs each of which is surrounded on its left and right edge by a region with the same linear dielectric medium.

To see this, begin by noting that the form of the solution in equations (9.1) and (9.2) is also found for a slab with general positions of the left and right edges of the slab along the x-axis. In this regard, only the $M_{i,j}$ are adjusted by the translation of the slab along the axis. This aspect of the slab solutions allows for a composition of the general solution of a layered medium in terms of products of matrices of the general form in (9.2).

Consider solving for the scattering of a plane wave normal incident from the left on a finite array of slabs, i.e., an array of N slabs. The solution begins at the right-hand edge of the layering, at the slab on the far right of the array. To the right of this slab only a transmitted wave is present so that for this particular slab $B_2 = 0$ is needed for the proper scattering boundary conditions. With this boundary condition (9.2) gives A_1 and A_2 immediately to the left of the slab at the right-hand end of the array in terms of B_1. Taking the resulting solutions A_1 and A_2 to be the B_1 and B_2 of the next slab to the left of the slab at the right-hand end of the array, the values of A_1 and A_2 to the left of that slab are computed from (9.2) and again used as the B_1 and B_2 of the next slab along the array. Proceeding in this way along the entire chain of the array, A_1 and A_2 for the incident and reflected waves at the far left of the array are obtained in terms of B_1 at the far right-hand end of the array. From the solutions at the two opposite ends of the chain the reflection coefficient is given by $R = |\frac{A_2}{A_1}|^2$ and the transmission coefficient by $T = |\frac{B_1}{A_1}|^2$ with the result that $R + T = 1$ for lossless systems. The process is the same for both linear and nonlinear systems.

For these types of linear and nonlinear medium one-dimensional systems a variety of problems can be handled [2, 4]. Many of the solutions have practical applications, but many offer solvable models which illustrate physical principles that come to play in a much more complex manner in higher-dimensional systems. Some examples of nonlinear properties discussed in layered media include: (a) band structure problems [4, 7] exhibiting renormalized band structures and gap soliton modes (b) impurity problems [4, 9], similar to those found in semiconductor electronics [10], in which the energy levels of the impurity bound states depend on the intensity of the fields at the impurity site (c) discussions of optical bistability and its effect on the transmission and dispersion properties of the modes of the system [8, 11, 12], and (d) various types of problems related to disorders in the one-dimensional array [13]. In addition, Kerr nonlinearity in one-dimensional systems has found technological applications in the development of optical switches [14] and optical diodes [15] for the design of opto-electric circuits. For the details of these treatments the reader is referred to the literature.

9.2 Nonlinearity in higher-dimensional photonic crystals and surface gratings

In higher-dimensional photonic crystals and in some surface grating problems, transfer matrix methods are no longer applicable [16, 17]. In particular, systems described by equations which cannot be reduced, in their simplest formulations, to one-dimensional equations must be handled by finite-difference time-domain methods or method of moments based treatments. In addition, for some Bragg grating problems [17] the method of coupled modes is used. This involves writing the solutions of the weakly nonlinear system as a linear combination of the modes of the linear limit of the system. The coefficients of the expansion are assumed to be slowly varying in space and time compared to the space and time variations of the modes and a set of coupled equations are generated for the coefficients. These methods, generally, are less accurate and require more work to program and implement on the more complex systems than are the transfer matrix methods applied to their respective simpler systems. The use of the finite-difference time-domain and method of moments to solve higher-dimensional nonlinear problems is a straightforward application of the methods described in an earlier chapter. Consequently, in the following only some references to typical results for photonic crystal systems obtained by these methods are mentioned. In the case of the study of split ring resonance (SRR)-nonlinear metamaterials the situation is much simpler than with photonic crystals. Most metamaterial problems have been handled by applying the Runga–Kutta method to the solution of the nonlinear differential equations describing the SRR array in such systems [18–21].

In most Bragg grating systems [17] problems have focused on the study of the dispersive properties and polarization effects, the treatment of gap solitons and their dispersion and velocity properties, self-transparence and problems involving localization of excitations, the modulational instabilities of excitations, the interaction of multiple pulses within these systems, and the treatment of nonlinear localized modes at impurity sites [16, 22]. Many of these lower-dimensional problems have been treated within the context of the coupled mode method [17]. In higher-dimensional photonic crystals containing waveguides and impurities the finite-difference time-domain and method of moments have been effective in treating the properties of these features. Some applications have been made in the study of optical transistors in photonic crystals [22, 23]. Studies in photonic crystals have also focused on the enhanced creation of slow light generated within these systems [24] and for the design of optical bistable switches [25]. These are typical types of simulation applications for nonlinear photonic crystals and photonic crystal circuits and address important issues in opto-electronic design.

The above discussions have focused on linear media systems and systems with Kerr nonlinearities. These conserve the frequency of the radiation. An additional, important nonlinear problem is that dealing with the generation of second harmonics of radiation from an applied fundamental frequency. We shall now discuss the generation of second harmonics. This will be followed by a generalization to parametric oscillation and amplification effects.

9.3 Generation of second harmonics

An important application of nonlinear optics is the use of nonlinearity to generate second harmonics of radiation [3, 26]. This has various applications in lasers [27, 28], where it is used to generate second harmonics from a laser output, and in optical diagnostics [29–31], where microscopy based on the ability of different materials to generated second harmonics is used to develop images of biological systems and surfaces.

An interesting application of photonic crystals is in enhancing the generation of second harmonics with the intent of offering reliable, intense, sources of such radiations. In particular, photonic crystals have provided solutions to problems of phase matching in the generation of second harmonics. This is a long known basic factor contributing to intensity loss in devices designed to generate second harmonic fields [26]. In the following, some elementary considerations of second harmonic generation will be presented and the phase matching problem briefly discussed. This will be followed by some discussion of the applications of photonic crystals to circumvent the phase matching problem.

To understand the simple ideas of second harmonic generation and phase matching, the equations for the generation of second harmonics in a uniform nonlinear medium will be outlined [26] and applied to the study of a uniform medium which generates a second harmonic in response to an applied radiation field. Consider Maxwell's curl equations in a uniform polarizable medium

$$\nabla \times \vec{E} = -\frac{\partial \vec{B}}{\partial t}, \tag{9.3a}$$

$$\nabla \times \vec{B} = \mu_0 \frac{\partial \vec{D}}{\partial t}, \tag{9.3b}$$

where $\vec{D} = \varepsilon_0 \vec{E} + \vec{P}$ so that

$$\nabla \times \vec{B} = \mu_0 \varepsilon_0 \frac{\partial \vec{E}}{\partial t} + \mu_0 \frac{\partial \vec{P}}{\partial t}. \tag{9.3c}$$

Here the polarization vector $\vec{P}$ describes the response of the medium to an applied field and contains both the linear and nonlinear terms. The nonlinearity of the medium which generates the second harmonic response enters specifically through $\vec{P}$ which also consists of a sum of terms for the linear contributions of the response that are well known from the study of standard electromagnetics. The linear terms in $\vec{P}$ are responsible for the renormalization of the permittivity of the medium at the fundamental frequency of the system, familiar from the study of the optics of linear media. The second harmonic generating terms introduce small fields at double the frequency of the fundamental applied field. This is the new element of the study here.

Consider the evaluation of (9.3c) for a fundamental frequency and the second harmonics generated by the nonlinear reaction of the system to the fundamental. The equation separates into two equations giving the behaviors of the fundamental and that of the generated field at double the fundamental frequency. From (9.3a) and (9.3c) the fields of the fundamental $\vec{E}_1$ at frequency ω satisfy

$$\nabla^2 \vec{E}_1 + \mu_0 \varepsilon(\omega) \omega^2 \vec{E}_1 = 0 \tag{9.4a}$$

where $\varepsilon(\omega)$ is renormalized by the linear part of the polarization, just as in the case when the nonlinearity is absent. The small generated second harmonic fields $\vec{E}_2$ at frequency 2ω satisfy the second equation

$$\nabla^2\vec{E}_2 + \mu_0\varepsilon(2\omega)(2\omega)^2\vec{E}_2 = -\mu_0(2\omega)^2\vec{P}_{2\omega}. \tag{9.4b}$$

Here $\vec{P}_{2\omega}$ is the nonlinear component of the polarization which generates the second harmonics and which is expressed in terms of the fundamental fields from which it derives. In (9.4b), the linear part of the polarization vector, $\vec{P}$, at frequency 2ω has been incorporated into the $\varepsilon(2\omega)$, renormalizing the free space permittivity at frequency 2ω, just as in the case when the nonlinearity is absent.

In a nonlinear system which generates a second harmonic a simple model for the nonlinear polarization is given by

$$\vec{P}_{2\omega} = \vec{E}_1 \cdot \tilde{d}(2\omega, \omega, \omega) \cdot \vec{E}_1 \tag{9.5}$$

where $\tilde{d}(2\omega,\omega,\omega)$ is a third rank tensor. This third rank tensor has many important symmetry properties [3, 26] which need to be considered for detailed treatments of second harmonic generation, and the reader is referred to the literature for these details. (A brief discussion of some of these symmetry properties is also given in a later section of this chapter.) Here a simplified treatment which illustrates the origins of the ideas of phase matching is the aim of the present discussion.

To understand the basic ideas of the problem of phase matching in second harmonic generation, a grossly simplified problem of propagation and generation in one-dimension will now be treated. The phase matching problem arises because, as the fundamental propagates through the nonlinear medium, it generates a second harmonic at each point in the medium. The second harmonics generated at each point add constructively and destructively through the media as they pass out of the source medium. Only under very special considerations do these constructive and destructive additions result in a total generated second harmonic wave of good intensity exiting the source device.

Consider a planar interface to the left of which is a vacuum and to the right of which is a nonlinear uniform medium with a second harmonic generating susceptibility (see figure 9.2 for a schematic). The x-axis is perpendicular to the planar interface which is located at $x = 0$ and an electromagnetic plane wave of frequency ω propagating along the x-axis is incident on the interface from the left. It is assumed that the generated second harmonic effect is very small so that the depletion of energy from the fundamental wave of frequency ω as it propagates into the nonlinear media is negligible. Within the nonlinear medium the solution from (9.4a) for the fundamental wave is of the form

$$E_1 = E_{10}e^{i[k(\omega)x-\omega t]} \tag{9.6}$$

where $k(\omega) = \sqrt{\mu_0\varepsilon(\omega)}\,\omega$ and the amplitude E_{10} is determined by the boundary conditions of the fundamental at the interface. The field E_{10} is in the plane perpendicular to the x-axis. In terms of the solution in (9.6), (9.4b) for the harmonic wave is written as

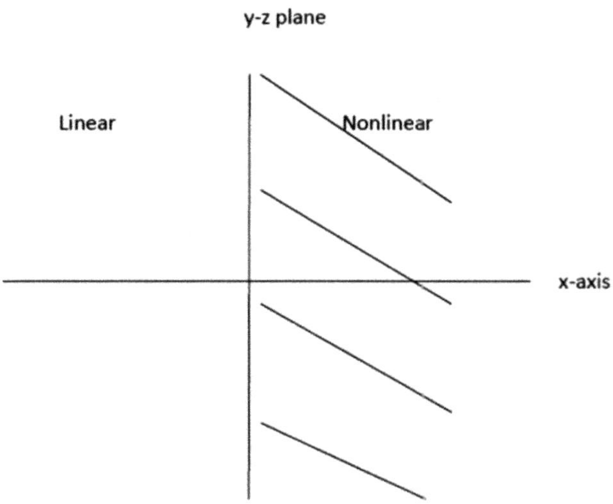

Figure 9.2. Planar interface between a linear medium and a nonlinear second harmonic generating medium.

$$\frac{\partial^2 E_2}{\partial x^2} + \mu_0 \varepsilon(2\omega)(2\omega)^2 E_2 = -\mu_0 (2\omega)^2 \mathrm{d}(2\omega)(E_{10})^2 \mathrm{e}^{2\mathrm{i}[k(\omega)x - \omega t]}. \tag{9.7}$$

Here the second harmonic generating polarization $P_{2\omega} = \mathrm{d}(2\omega)(E_{10})^2 \mathrm{e}^{2\mathrm{i}[k(\omega)x-\omega t]}$ is in the plane perpendicular to the x-axis and parallel to E_2. The polarizations of the incident and generated waves are assumed to be consistent with the microscopic symmetry of the materials so that the coefficient of the polarization plane wave is constant in space [26].

To obtain a solution of (9.7) E_2 is assumed to be of the form

$$E_2 = E_{20}(x)\mathrm{e}^{\mathrm{i}[k(2\omega)x - 2\omega t]} \tag{9.8}$$

where $k(2\omega) = \sqrt{\mu_0 \varepsilon(2\omega)}\, 2\omega$ and $E_{20}(x)$ is assumed to be a slowly varying envelope function in comparison to the plane wave which it multiplies. In addition, it is assumed that the envelope function changes slowly enough in space that it satisfies $| \frac{\partial^2 E_{20}}{\partial x^2} | \ll | k(\omega)\frac{\partial E_{20}}{\partial x} |$. Under these conditions

$$\frac{\partial^2 E_2}{\partial x^2} \approx \left[2\mathrm{i}k(2\omega)\frac{\partial E_{20}}{\partial x} - k^2(2\omega)E_{20} \right] \mathrm{e}^{\mathrm{i}[k(2\omega) - 2\omega t]}, \tag{9.9}$$

and from (9.7) we find that

$$\frac{\partial E_{20}}{\partial x} = -\frac{\mu_0 2\omega \mathrm{d}(2\omega)(E_{10})^2}{2\mathrm{i}\sqrt{\mu_0 \varepsilon(2\omega)}} \mathrm{e}^{\mathrm{i}[2k(\omega) - k(2\omega)]x}. \tag{9.10}$$

The amplitude of the electric field is obtained by integrating (9.10) subject to the boundary condition that $E_{20}(x = 0) = 0$, i.e., the generated wave begins to be generated at the surface of the semi-infinite nonlinear medium as the fundamental begins to pass into the nonlinear medium. Phase matching occurs under the very specialized condition that $2k(\omega) = k(2\omega)$. Solving (9.10) for $x>0$ with this condition

$$E_{20}(x) = -\frac{\mu_0 2\omega \mathrm{d}(2\omega)(E_{10})^2}{2\mathrm{i}\sqrt{\mu_0 \varepsilon(2\omega)}} x, \tag{9.11}$$

and the field amplitude grows linearly with the distance from the interface. For $2k(\omega) \neq k(2\omega)$, the phase miss-match condition, an integration of (9.10) leads to a bounded oscillating field amplitude. Clearly, the preferred condition for generating an intense second harmonic source output is the linear growth with position obtained in (9.11) at the condition of phase matching. In general it is difficult to satisfy the phase matching condition as it requires a very precise set of conditions places on the dielectric properties and dispersion relations of the fundamental and second harmonics within the nonlinear system.

Metamaterials [32] and photonic crystals [33–35] offer two different types of methods for enhancing the generation of second harmonics. Metamaterial mirrors have recently been developed [32] which afford great enhancements of the second harmonics that are generated at the mirror surface. In these mirrors the wavelengths of the fundamental and second harmonic radiations are greater than the length scales of the nanostructures forming the mirrored surface, and this is the reason for their classification as metamaterials. However, the function of the mirror is by the mechanisms of nanostructure quantum wells [26] rather than by the nanostructure SRRs of our earlier treatments of SRR metamaterials. The mirror surface is composed of a multi-quantum-well semiconductor heterostructure on a substrate of metal (gold) and has a periodic metal (gold) patterning of facets on the reflecting side of the mirror surface, opposite the substrate. The fundamental frequency of light is incident normal to the mirror surface and its second harmonic is generated in the reflected normal direction. The enhanced nonlinearity is facilitated by the fields of the surface plasmons excited in the heterostructure forming the mirror surface.

Photonic crystals [33–35], on the other hand, facilitate the generation of second harmonics through effects related to their periodic properties. In these systems the wavelengths of the fundamental and/or second harmonics are of the order of those of the direct lattice vectors of the photonic crystals. The operating mechanisms of the photonic crystals for second harmonic studies arise from: (a) the periodic poling of materials, creating a periodic nonlinear response, (b) the design of a periodic dielectric in which the fundamental and/or second harmonic interact with the band structure of the photonic crystal, or (c) using the photonic crystal to modify the density of states via the Purcell effect.

Two basic ideas are operative in the application of photonic crystals. In the first, consider the example treated earlier of the uniform semi-infinite slab and its problem of phase matching. A rough solution of the problem can be made by making a finite array of second harmonic generating slabs. This is done by slicing the uniform semi-infinite slab up along the y–z plane and removing slabs so that only a finite number of slabs remain. The remaining slabs are chosen so that their generated fields add constructively in the remaining array. The remaining slabs should represent an improvement over the second harmonic generated by a bulk system formed by filling the spaces between neighboring slabs with the second harmonic generating medium. The filling in the bulk system adds destructively to the fields of the chosen array. A finite periodic array is easily formulated to meet the above conditions [33–35].

In the second approach the ability of photonic crystals to modify the electromagnetic density of states is employed. This is the Purcell effect. Transition processes obtained from the Fermi golden rule depend on the electromagnetic density of states so that enhancing the density of states increases transition rates. Specifically, immediately above and below the stop bands of a photonic crystal the density of states are increased over the density of states in a uniform medium. The idea of the second method is to arrange for a system in which the enhancement of the density of states at both the fundamental and harmonic frequency enhances the generation of second harmonics over that which would occur generated in a uniform medium [33–35].

9.4 Some symmetry properties of the second harmonic tensor

The tensor in (9.5) for the frequency 2ω polarization vector generated by an application of a frequency ω electric field has symmetry properties which greatly restrict the crystallography of the generating material. Specifically, a consideration of the spatial properties of the tensor components of $\tilde{d}(2\omega, \omega, \omega)$ under a space inversion leads to the conclusion that only materials which are not invariant under spatial inversion (i.e., an operation represented by the transformation $\vec{r} \rightarrow -\vec{r}$.) can support the generation of second harmonics of radiation. This is a significant and fundamental limitation on schemes for second harmonic generation.

To see this symmetry property [26], consider the representation of the second harmonic tensor $\tilde{d}(2\omega, \omega, \omega)$ in the Cartesian coordinates x_1, x_2, x_3. In this representation the components of the tensor are of the form $d_{i,j,k}$ where i, j, k run over the integers 1, 2, 3 of the three orthogonal axes. Written in the context of these coordinates, the polarization relationship in (9.5) takes the general form [26]

$$P_{2\omega,i} = \sum_{j,k} d_{i,j,k} E_{1,j} E_{1,k} \tag{9.12}$$

where $E_{1,j}$ and $E_{1,k}$ are components of the ω applied frequency fields. From (9.12) a fundamental limitation on the $d_{i,j,k}$ components is now obtained.

Specifically, under a spatial inversion of the physical system $P_{2\omega,i} \rightarrow -P_{2\omega,i}$ and $E_{1,j} \rightarrow -E_{1,j}$ so that if $d_{i,j,k} \rightarrow d_{i,j,k}$ is invariant under the inversion, it follows that [26]

$$-P_{2\omega,i} = \sum_{j,k} d_{i,j,k}(-E_{1,j})(-E_{1,k}) = P_{2\omega,i} = 0. \tag{9.13}$$

Consequently, the only way that (9.13) can be valid for a nonzero $P_{2\omega,i}$ is if $d_{i,j,k}$ changes under the inversion. This provides a fundamental limitation on the crystallography of the material, and, in particular, eliminates such common materials as water, glass, and systems with cubic symmetry from the considerations of second harmonic applications.

Table 9.1. Examples of some second harmonic generating materials [26, 50–52].

Inorganic	Quartz crystals	KH_2PO_4	$Ba_2NaNb_5O_{15}$
Elements	*Te*	*Se*	
Organics and biologicals	TCNQ derivatives	Perhyrrotriphenylene	Stilbazolium dyes

In table 9.1 some examples of materials exhibiting second harmonic generation are presented. These are only a few representative examples and many more are available in the literature. In addition, important sources of second harmonic generation occur at various geometric features such as surfaces, slabs, and roughness or in disordered systems. These later systems introduce symmetry breaking components, often at a macroscopic level, and the second harmonics generated by such features have also found important applications in the study of roughness and disorder and in studies of biological surfaces.

9.5 Parametric oscillation and amplification: frequency sums

The discussions for second harmonic generation can be generalized to consider materials exhibiting parametric oscillation and amplification [26]. In our earlier considerations of second harmonic generation, the frequency of radiation ω is doubled so that $\omega + \omega = 2\omega$ where the sum is generated in the nonlinearity of the medium. Similarly, in parametric oscillation and amplification, three frequencies of radiation ω_1, ω_2, ω_3 are involved so that $\omega_1 + \omega_2 = \omega_3$ where the frequency sum ω_3 is again generated in the nonlinearity of the medium. Consequently, the generation of second harmonics is a special case of the $\omega_1 + \omega_2 = \omega_3$ summation. In addition, the nonlinear processes can also proceed in the reverse direction so that a single frequency component splits up into two lower frequency components.

For the treatment of the parametric phenomena, the polarization tensor equation in (9.12) is generalized to take the form

$$P_{3,i} = \sum_{j,k} d_{i,j,k}(\omega_3, \omega_2, \omega_1)E_{2,j}E_{1,k}, \tag{9.14}$$

where the subscripts 1, 2, 3 indicate fields and polarizations with frequency components ω_1, ω_2, ω_3, respectively. In particular, for the following considerations (9.14) is evaluated for plane waves given by $P_{3,i} = P_{3,i,0}e^{i\left(\vec{k}_3\cdot\vec{r}-\omega_3 t\right)}$, $E_{2,j} = E_{2,j,0}e^{i\left(\vec{k}_2\cdot\vec{r}-\omega_2 t\right)}$, $E_{1,k} = E_{1,k,0}e^{i\left(\vec{k}_1\cdot\vec{r}-\omega_1 t\right)}$ where the frequencies are found to add yielding $\omega_1 + \omega_2 = \omega_3$. Reduced in the appropriate limit, (9.14) yields the nonlinear polarization treated in the earlier discussions of second harmonic generation. From (9.14) it is also seen that in considering the sum of wavevectors there is a similar phase matching problem involving the wavevectors as that found in second harmonic generation. Consequently, treatments of phase matching are included in some of our later discussions of parametric solutions.

The tensor notation in (9.14) becomes quite cumbersome in discussing parametric phenomena so that in the following a simpler model [26] is introduced which is easier to treat in developing an understanding of parametric phenomena. This model portrays the gist of the physics without the reader becoming bogged down in the complexities of notational bookkeeping.

9.5.1 A simplified notation for a simplified model

In order to give a simplified discussion modeling the behaviors found in parametric phenomena of nonlinear media, the components of the polarization are assumed to be of a simple form. As a first simplification the electric fields and polarizations are chosen to all be parallel to one another, and the waves in the media are chosen to be plane waves propagating along the x-direction. Consequently, the theory reduces to a study of a scalar formulation of the wave equation in one-dimension [26].

As a second simplification the nonlinear components of the polarization are assumed to be related to the electric field components by the form [26]

$$P_N(x, t) = d[E_{\omega_1}(x, t) + E_{\omega_2}(x, t) + E_{\omega_3}(x, t)]^2 \tag{9.15}$$

where now

$$E_{\omega_1}(x, t) = \frac{1}{2}[E_1(x)e^{i(k_1x-\omega_1t)} + c.\ c] \tag{9.16a}$$

$$E_{\omega_2}(x, t) = \frac{1}{2}[E_2(x)e^{i(k_2x-\omega_2t)} + c.\ c] \tag{9.16b}$$

$$E_{\omega_3}(x, t) = \frac{1}{2}[E_3(x)e^{i(k_3x-\omega_3t)} + c.\ c] \tag{9.16c}$$

are the electric fields at the three frequencies of interest. In the plane waves of (9.16) the linear dispersion relations, relating the wavevectors and the frequencies of the waves, have the forms

$$k_1^2 = \omega_1^2\mu_0\varepsilon_1, \quad k_2^2 = \omega_2^2\mu_0\varepsilon_2, \quad k_3^2 = \omega_3^2\mu_0\varepsilon_3 \tag{9.17}$$

where $\varepsilon_1 = \varepsilon_0(1 + \chi_1)$, $\varepsilon_2 = \varepsilon_0(1 + \chi_2)$, $\varepsilon_3 = \varepsilon_0(1 + \chi_3)$ are the linear (i.e., zero field limit) parts of the dielectric constants of the medium at the three different frequencies ω_1, ω_2, ω_3 and χ_1, χ_2, χ_3 are the linear parts of the electric susceptibilities at these same frequencies. In addition, it is important to note that in (9.15) the coefficient d of the nonlinearity is a scalar constant, and for the discussions only the polarization at the frequencies ω_1, ω_2, ω_3 is of interest. The other frequencies generated in (9.15) will not enter the following treatments.

To begin, consider the system in the absence of the nonlinear polarization. The wave equations for the three noninteracting frequency components ω_1, ω_2, ω_3 of the radiation are given by the linear wave equations

$$\frac{\partial^2}{\partial x^2}E_{\omega_j}(x, t) = \mu_0\sigma_j\frac{\partial}{\partial t}E_{\omega_j}(x, t) + \mu_0\varepsilon_j\frac{\partial^2}{\partial t^2}E_{\omega_j}(x, t) \tag{9.18}$$

where $j = 1, 2, 3$ runs over the three frequencies. This is the general form of a wave equation for a linear medium in which the system is allowed to have a conductivity σ_j.

Next, let us consider the propagation through the system of the plane wave components in (9.16) which are of the form $E_j(x)e^{i(k_jx-\omega_jt)}$. The motion of the other $E_j^*(x)e^{-i(k_jx-\omega_jt)}$ are obtained directly from these. This allows us to use plane wave Fourier components to reduce the number and complexity of the equations generated in the solution of the wave equations.

From (9.16) and (9.18) for each frequency component of the plane waves it then follows that

$$\frac{\partial^2}{\partial x^2}\left[E_j(x)e^{ik_jx}\right] = \left[-i\omega_j\mu_0\sigma_j - \omega_j^2\mu_0\varepsilon_j\right]E_j(x)e^{ik_jx}. \tag{9.19}$$

Taking the derivatives of the term on the left of (9.18) then gives the operator form [26]

$$\begin{aligned}\frac{\partial^2}{\partial x^2}\left[E_j(x)e^{ik_jx}\right] &= -\left[k_j^2E_j(x) - 2ik_j\frac{d}{dx}E_j(x) - \frac{d^2}{dx^2}E_j(x)\right]e^{ik_jx}\\ &\approx -\left[k_j^2E_j(x) - 2ik_j\frac{d}{dx}E_j(x)\right]e^{ik_jx}\end{aligned} \tag{9.20}$$

where here in the last approximation on the right side it is assumed that $\left|k_j\frac{d}{dx}E_j(x)\right| \gg \left|\frac{d^2}{dx^2}E_j(x)\right|$. This approximation assumes the e^{ik_jx} has a much more rapid variation in space than does $E_j(x)$ and is often referred to as an envelope approximation [26]. In this regard, it is being assumed that the nonlinear effects are small so that the change in amplitude of the electric fields occur slowly on the scale of the wavelength of the radiation.

To arrive at the full wave equation including nonlinear effects it only remains to include the nonlinear polarization in the problem. This greatly increases the difficulty of the problem.

To account for the nonlinear interactions, the jth frequency component of the nonlinear polarization from (9.15) must be added to the right side of (9.19). For example, treating (9.15) for the case that $j = 3$ in the consideration of generated $\omega_3 = \omega_1 + \omega_2$ radiation, the nonlinear polarization obtained from (9.15) is of the form

$$P_{N,3} = \frac{1}{2}dE_1(x)e^{i(k_1x-\omega_1t)}E_2(x)e^{i(k_2x-\omega_2t)} = \frac{1}{2}dE_1(x)E_2(x)e^{i(k_1+k_2)x}e^{-i\omega_3t}. \tag{9.21a}$$

Here the subscript 3 on $P_{N,3}$ indicates that it is the component of P_N with the $e^{-i\omega_3t}$ time dependence. Similar results when considering $j = 1$ or $j = 2$ follow for

$$P_{N,1} = \frac{1}{2}dE_3(x)E_2^*(x)e^{i(k_3-k_2)x}e^{-i\omega_1t} \tag{9.21b}$$

and

$$P_{N,2} = \frac{1}{2}dE_3(x)E_1^*(x)e^{i(k_3-k_1)x}e^{-i\omega_2t} \tag{9.21c}$$

frequency components of the polarization.

With the introduction of the nonlinear polarization into (9.19) and (9.20) these three equations become coupled by the ω_1, ω_2, ω_3 frequency components of the of the polarization in (9.15). The resulting couplings come from the mixing of the frequency components in the squared term in (9.15) as shown in (9.21). Specifically, to the right-hand side of (9.18) through (9.20) is added the term [26]

$$\mu_0 \frac{\partial}{\partial t^2} P_{N,j}. \tag{9.22}$$

From this addition follows the three coupled dynamical equations for the electric fields at frequencies ω_1, ω_2, ω_3 [26]

$$\frac{d}{dx}E_1(x) = -\frac{\sigma_1}{2}\sqrt{\frac{\mu_0}{\varepsilon_1}}E_1(x) + i\frac{\omega_1}{2}\sqrt{\frac{\mu_0}{\varepsilon_1}}dE_3(x)E_2^*(x)e^{i(k_3-k_2-k_1)x} \tag{9.23a}$$

$$\frac{d}{dx}E_2^*(x) = -\frac{\sigma_2}{2}\sqrt{\frac{\mu_0}{\varepsilon_2}}E_2^*(x) - i\frac{\omega_2}{2}\sqrt{\frac{\mu_0}{\varepsilon_2}}dE_1(x)E_3^*(x)e^{-i(k_3-k_2-k_1)x} \tag{9.23b}$$

$$\frac{d}{dx}E_3(x) = -\frac{\sigma_3}{2}\sqrt{\frac{\mu_0}{\varepsilon_3}}E_3(x) + i\frac{\omega_3}{2}\sqrt{\frac{\mu_0}{\varepsilon_3}}dE_1(x)E_2(x)e^{-i(k_3-k_2-k_1)x}. \tag{9.23c}$$

The next stage is to discuss some example solutions of the system in (9.23).

9.5.2 Examples of solutions

In this section some representative examples of the evaluation of (9.23) are given for simple problems involving parametric amplification and oscillation [26]. To provide a simplification of the treatment in the evaluation of the system in (9.23), it is assumed that of the three different frequency components in the system one of the electric fields is much larger than the other two. In addition, the largest field component is regarded as a position independent constant. This restricts the spatial variation of field intensities along the x-direction to the other two frequency component fields and provides a model which can be treated analytically.

To begin, it is helpful to rewrite (9.23) in a less cluttered notation. In this regard, a notation is adopted in which the fields and the system parameters are redefined by renormalized forms which are now given. Specifically, the fields are rewritten as

$$A_l = \sqrt{\frac{n_l}{\omega_l}}E_l, \text{ for } l = 1, 2, 3 \tag{9.24}$$

where n_l is the index of refraction in the zero field limit, the nonlinear coupling parameter is given by the form

$$D = d\sqrt{\frac{\mu_0}{\varepsilon_0}\frac{\omega_1\omega_2\omega_3}{n_1n_2n_3}}, \tag{9.25}$$

and the dissipation in the system is characterized by

$$\alpha_l = \sigma_l \sqrt{\frac{\mu_0}{\varepsilon_l}}, \text{ for} = 1, 2, 3. \tag{9.26}$$

Within this notation (9.23) then become the set of coupled nonlinear equations [26]

$$\frac{d}{dx}A_1(x) = -\frac{1}{2}\alpha_1 A_1(x) + i\frac{1}{2}DA_3(x)A_2^*(x)e^{i\Delta kx} \tag{9.27a}$$

$$\frac{d}{dx}A_2^*(x) = -\frac{1}{2}\alpha_2 A_2^*(x) - i\frac{1}{2}DA_1(x)A_3^*(x)e^{-i\Delta kx} \tag{9.27b}$$

$$\frac{d}{dx}A_3(x) = -\frac{1}{2}\alpha_3 A_3(x) + i\frac{1}{2}DA_1(x)A_2(x)e^{-i\Delta kx} \tag{9.27c}$$

where $\Delta k = k_3 - k_2 - k_1$ characterizes the phase mismatch in the system.

9.5.2.1 First example solution

The first interesting solution of the set of equations in (9.27) is obtained for the case in which $\alpha_1 = \alpha_2 = \alpha_3 = 0$ so that there are no dissipative losses in the system and for initial conditions on $A_1(0), A_2(0), A_3(0)$ at $x = 0$ such that $\omega_1 \mid A_1(0) \mid^2, \omega_2 \mid A_2(0) \mid^2 \ll \omega_3 \mid A_3(0) \mid^2$. This last condition is a statement that the energy current ($\propto \hbar\omega \times$ field intensity) carried in the $A_3(0)$ field is much greater than that in the other fields. The study of (9.27) then reduces to a treatment of the two equations [26]

$$\frac{d}{dx}A_1(x) = i\frac{1}{2}DA_3(0)A_2^*(x)e^{i\Delta kx} = i\frac{1}{2}gA_2^*(x)e^{i\Delta kx} \tag{9.28a}$$

$$\frac{d}{dx}A_2^*(x) = -i\frac{1}{2}DA_1(x)A_3^*(0)e^{-i\Delta kx} = -i\frac{1}{2}gA_1(x)e^{-i\Delta kx} \tag{9.28b}$$

where $g = DA_3(0)$ and it is assumed that $A_3(0)$ is real. Here in line with our earlier remarks the field $A_3(x) = A_3(0)$ is regarded as a constant.

The solution of (9.28) leads directly to the study of parametric amplification which is now addressed.

9.5.2.2 Parametric amplification

First consider the property of parametric amplification. This arises directly from the solution of the problem posed earlier in (9.28), and for simplicity only the phase matching case of the solution (for which $\Delta k = 0$) is treated.

Specifically, under these conditions the general form of the solution of the system in (9.28) is given by [26]

$$A_1(x) = A_1(0)\cosh\frac{g}{2}x + iA_2^*(0)\sinh\frac{g}{2}x \tag{9.29a}$$

$$A_2^*(x) = A_2^*(0)\cosh\frac{g}{2}x - iA_1(0)\sinh\frac{g}{2}x. \tag{9.29b}$$

For the initial condition $A_2^*(0) = 0$ this reduces to

$$A_1(x) = A_1(0)\cosh\frac{g}{2}x \tag{9.30a}$$

$$A_2^*(x) = -iA_1(0)\sinh\frac{g}{2}x, \tag{9.30b}$$

and it is found that both of the ω_1 and ω_2 frequency components increase exponential in the distance traveled by the wave. The increase in the amplitudes of these components is drawn from their interaction with the high intensity $\omega_3 > \omega_1$, ω_2 frequency component. In this regard, the ω_1, ω_2 components are amplified as they travel through space by their interactions mediated by the nonlinearity of the medium.

In the case that $\Delta k \neq 0$ the integration of (9.28) is a little more difficult. The results of this integration give rise to constructive and destructive interference effects associated with the problem of phase matching.

9.5.2.3 Second example: parametric oscillations

A closely related problem to the study of parametric amplification is that of parametric oscillation. Typical parametric oscillation problems involve the study of a lossy medium enclosed within a resonant cavity, and of interest are the different regions and types of solutions involved with propagation within the cavity. Of particular interest are waves propagating at the cavity resonant frequencies.

Consider now adding back into (9.27) the loss components. The resulting set of equations then become [26]

$$\frac{d}{dx}A_1(x) = -\frac{1}{2}\alpha_1 A_1(x) + i\frac{1}{2}gA_2^*(x)e^{i\Delta kx} \tag{9.31a}$$

$$\frac{d}{dx}A_2^*(x) = -\frac{1}{2}\alpha_2 A_2^*(x) - i\frac{1}{2}gA_1(x)e^{-i\Delta kx}. \tag{9.31b}$$

As in the case of parametric amplification only the simplest case in which $\Delta k = 0$ is considered in the following treatment. This avoids phase matching problems that are a mathematical complication and which have standard methods of resolution [26].

An interesting case of (9.31) is found when the medium is placed within a resonance cavity formed of reflecting infinite parallel plates perpendicular to the x-axis. Now under the proper conditions some of the electromagnetic fields propagating along the x-axis travel indefinitely between the cavity walls. This follows as the system in the cavity has both generative and dissipative components which if correctly balanced allow for a steady state condition to be achieved within the cavity.

Specifically, for a fixed $A_3(x) = A_3(0)$ a steady state or indefinite state of motion in space can be sustained between the plates. This occurs under the condition in (9.31) that $\frac{d}{dx}A_1(x) = \frac{d}{dx}A_2^*(x) = 0$, requiring the determinate of the coefficient matrix in (9.31),

$$\begin{vmatrix} -\frac{1}{2}\alpha_1 & i\frac{1}{2}g \\ -i\frac{1}{2}g & -\frac{1}{2}\alpha_2 \end{vmatrix}, \tag{9.32}$$

to becomes zero. For this condition to be fulfilled it is needed that

$$g^2 = \alpha_1\alpha_2 \tag{9.33}$$

which is the condition balancing the nonlinear generation of the fields with their dissipation. In the case that $\Delta k \neq 0$ the integration of (9.31) is more difficult due to phase match problems.

Next an important case of (9.27) known as frequency up-conversion is treated. This is closely related to second harmonic generation.

9.5.2.4 Third example: frequency up-conversion

Frequency up-conversion is a type of generalization of second harmonic generation. For this case of (9.27) it is assumed that $\omega_1 \mid A_1(0) \mid^2$, $\omega_3 \mid A_3(0) \mid^2 \ll \omega_2 \mid A_2(0) \mid^2$ and now $A_2(x) = A_2(0)$ is constant in space. In addition, for a simplification, dissipative effects are ignored so that $\alpha_1 = \alpha_2 = \alpha_3 = 0$. The focus in the following study [26] is then to treat the case in which the ω_1 and ω_2 fields are applied to a nonlinear medium which then generates an output field at ω_3.

Under the conditions just given it follows from (9.27) that [26]

$$\frac{d}{dx}A_1(x) = i\frac{1}{2}gA_3(x)e^{i\Delta kx} \tag{9.34a}$$

$$\frac{d}{dx}A_3(x) = i\frac{1}{2}gA_1(x)e^{-i\Delta kx} \tag{9.34b}$$

where $g = \sqrt{\frac{\mu_0}{\varepsilon_0}\frac{\omega_1\omega_3}{n_1n_3}}dE_2$, $A_2(0)$ is assumed to be real, and the problem is again to be treated in the limit that $\Delta k = 0$. This eliminates the necessity of phase matching considerations.

With these conditions, the general solution of (9.34) is obtained as

$$A_1(x) = A_1(0)\cos\frac{g}{2}x + iA_3(0)\sin\frac{g}{2}x \tag{9.35a}$$

$$A_3(x) = A_3(0)\cos\frac{g}{2}x + iA_1(0)\sin\frac{g}{2}x. \tag{9.35b}$$

Initially only the ω_1 and ω_2 fields are applied to the medium so that for $A_3(0) = 0$ the solutions in (9.35) take the particular form

$$A_1(x) = A_1(0)\cos\frac{g}{2}x \tag{9.36a}$$

$$A_3(x) = iA_1(0)\sin\frac{g}{2}x. \tag{9.36b}$$

The weak ω_1 applied field and ω_3 generated field in (9.36) are found to oscillate in space with a ninety-degree phase difference. In this regard, the ω_2 field is constant in amplitude and acts with the nonlinearity to shuttle energy between the two other frequency components.

Notice also that the generated frequency ω_3 field is at a higher frequency than are ω_1 and ω_2 and this is the origin of the term 'up-conversion'. In addition, from (9.36) it is seen that [26]

$$| A_1(x) |^2 + | A_3(x) |^2 = | A_1(0) |^2 \tag{9.37}$$

correlates the spatial variation of the various frequency components.

The result in (9.36) specifically show that, if a strong field of frequency ω_2 and a weak field of frequency ω_1 are applied to the medium, a second weak field of frequency $\omega_3 = \omega_1 + \omega_2$ is generated in the nonlinear medium. Second harmonic generation is the special case of the up-conversion results in which $\omega_1 = \omega_2$.

9.5.2.5 Four wave mixing

Another process of interest is that of four wave mixing [53–59]. This is essentially a higher order nonlinear interaction than those studied to this point and involves the simultaneous interaction of four photons with an energy conservation scheme of the form $\omega_1 + \omega_2 = \omega_3 + \omega_4$. As a consequence, it is generally a much weaker effect that those discussed earlier.

Associated with the energy conservation of these processes there is also a phase matching problem associated with the study of four wave mixing. This is an added difficulty in the observation of these processes. Nevertheless, four wave mixing has been of some recent interest.

9.5.3 Examples of systems: optical fibers

An important set of systems in which the variety of nonlinear processes discussed earlier occur are optical fibers [53–56]. Optical fibers are long fiber strands of dielectric which are used as waveguides to transport electromagnetic energy and information. Of recent they have been developed using ideas of photonic crystal technology in which periodicity of the dielectric has been introduced in the plane perpendicular to the axis of the fiber [53, 56]. The purpose of the periodicity falls into two categories. In the first type of application the periodicity acts to form a composite material and is used to spatially modulate the dielectric properties of the fiber over the fiber cross section. In this way the effective dielectric properties of the fiber are continuously varied radially. This develops a type of cladding which aids in binding traveling electromagnetic waves to the waveguide. In the second type of application the periodicity acts as a true photonic crystal, opening stop band frequency gaps in the electromagnetic spectrum for motion of the guided radiation in the radial direction. These stop bands are again used to confine the radiation to propagate along the fiber axis.

In the following some introductory discussions of the nonlinear properties of fiber optic waveguides are treated within the context of a basic model [53–56]. These illustrate the properties of nonlinear effects in fiber optics waveguides. A simple model is initially introduced, followed by a presentation of its basic solutions.

For a first consideration a Kerr nonlinear medium is introduced into the waveguide, and a discussion of the soliton solutions and their propagation along the fiber waveguide is given. This is followed by a generalization to systems exhibiting parametric oscillations and amplification.

9.5.3.1 Simple model

Consider the solutions of a waveguide composed as an infinite cylinder of a homogeneous isotropic medium of uniform dielectric constant. For the case of a linear dielectric medium the guided modes of the waveguide structure are solved in standard presentations [57, 58]. In these presentations it is shown for a dielectric cylinder with an axis along the z-axis of a cylinder coordinate system that the electric and magnetic fields of the guided modes have the form [57, 58]

$$\vec{E}(r, \theta, z) = \vec{E}_0(r, \theta)e^{i(kz-\omega t)} \tag{9.38a}$$

$$\vec{H}(r, \theta, z) = \vec{H}_0(r, \theta)e^{i(kz-\omega t)} \tag{9.38a}$$

with modal dispersion relations given by [57, 58]

$$k_\alpha = \sqrt{\mu\varepsilon}\sqrt{\omega^2 - \omega_\alpha^2} \tag{9.39}$$

where $\alpha = 1, 2, 3, \ldots.$ is a mode index with $\omega_1 < \omega_2 < \omega_3 < \ldots.$ As is shown in the following, these results can now be directly generalized to treat the properties of systems with weak dielectric nonlinearity.

In this regard, in the following a simple model of the waveguide in the presence of a weak nonlinearity is developed directly by considering the dispersion relation in (9.39) for a waveguide carrying a single mode. This is done from a consideration of the general form $k = \sqrt{\mu\varepsilon}\sqrt{\omega^2 - \omega_a^2}$ of the dispersion relation in (9.39) for a mode with cutoff frequency ω_a, and under the assumption that the nonlinearity enters as a weak perturbation of the system. We begin with a treatment of Kerr media fibers and their soliton solutions.

9.5.3.2 Soliton modes in a Kerr medium fiber

Begin by Taylor series expanding $k(\omega) = \sqrt{\mu\varepsilon}\sqrt{\omega^2 - \omega_a^2}$ about a wavevector and frequency ω_0, k_0 satisfying $k_0 = \sqrt{\mu\varepsilon}\sqrt{\omega_0^2 - \omega_a^2}$. This expansion gives the approximate relation [53–56]

$$k - k_0 = k'(\omega_0)(\omega - \omega_0) + \frac{1}{2}k''(\omega_0)(\omega - \omega_0)^2 \tag{9.40}$$

which is valid in the neighborhood of ω_0, k_0. The question now is how, to lowest order in the electric field, does the expansion in (9.40) change in the presence of an electric field intensity. To understand the change, the properties of the medium in the presence of a weak Kerr nonlinear dielectric are first considered.

When a small Kerr nonlinearity is added into the problem the dielectric constant of the system takes the form

$$n(\omega) = n_0(\omega) + n_1 \mid \vec{E} \mid^2 \tag{9.41a}$$

where the first term on the right is the linear (zero field) limit of the dielectric function and the second term on the right is the Kerr nonlinear contribution. The Kerr contribution depends on the amplitude squared of the electric field, $\mid \vec{E} \mid^2$, and as a result of this generalization of the dielectric function the system wavevector $k(\omega)$ is now going to also depend on $\mid \vec{E} \mid^2$. Consequently, the dispersion relation for the modified system becomes of the general form

$$k - k_0 = f\left(\omega, \mid \vec{E} \mid^2\right) \tag{9.41b}$$

which reduces to (9.40) in the limit that $\mid \vec{E} \mid^2 = 0$.

Applying a Taylor series to the new dispersion relation in (9.41b) gives an expansion for the dispersion about k_0, ω_0, $\mid \vec{E} \mid^2 = 0$ yielding [53–56]

$$\begin{aligned} k - k_0 &= k'(\omega_0)(\omega - \omega_0) + \frac{1}{2}k''(\omega_0)(\omega - \omega_0)^2 + \left.\frac{\partial f\left(\omega_0, \mid \vec{E} \mid^2\right)}{\partial \mid \vec{E} \mid^2}\right|_{\mid \vec{E} \mid^2 = 0} \mid \vec{E} \mid^2. \\ &= k'(\omega_0)(\omega - \omega_0) + \frac{1}{2}k''(\omega_0)(\omega - \omega_0)^2 + c \mid \vec{E} \mid^2. \end{aligned} \tag{9.42}$$

The dispersion relation in (9.42) can now be cast into the form of a differential equation for an electric field envelope function describing the guided mode. This is accomplished by introducing some assumptions based on the fact that in the absence of nonlinearity the z-dependence of the modal wave functions are plane waves. Specifically, assuming a weak nonlinearity we first replace some of the variables in (9.42) with differential operators by making the following associations:

$$k - k_0 \leftrightarrow -i\frac{\partial}{\partial z} \tag{9.43a}$$

$$\omega - \omega_0 \leftrightarrow i\frac{\partial}{\partial t}. \tag{9.43b}$$

Next, we assume an envelope function $q(z, t)$ exists which modulates the plane wave form $e^{i[k_0 z - \omega_0 t]}$ of the electric field mode at wavevector k_0 and frequency ω_0. In this way the envelope function satisfies

$$(k - k_0)q(z, t) = -i\frac{\partial}{\partial z}q(z, t) \tag{9.43c}$$

$$(\omega - \omega_0)q(z, t) = i\frac{\partial}{\partial t}q(z, t) \tag{9.43d}$$

where it is seen that $q(z, t) = q_0 e^{i[(k-k_0)z-(\ \omega-\omega_0)t]}$ satisfies both of (9.43c) and (9.43b) exactly. With these associations (9.42) takes the form of a differential wave equation for a general envelope function $q(z, t)$ given by

$$i\left(\frac{\partial q}{\partial z} + k'\frac{\partial q}{\partial t}\right) - \frac{1}{2}k''\frac{\partial^2 q}{\partial t^2} + c \mid q \mid^2 q = 0 \tag{9.44}$$

where $\mid q \mid^2 = \mid \vec{E} \mid^2$.

Here $q(z, t)$ is an indicator of the amplitude of the electric field and as such performs the role of an envelope function for the field [53–58]. In this regard, $q(z, t)$ could be defined based on an average of the field over some radial region at z or some other reasonable indicator of the behavior of the field at that point along the waveguide axis.

Notice that in the linear limit in which $c = 0$, the solution

$$q(z, t) = q_0 e^{i[(k-k_0)z-(\omega-\omega_0)t]} \tag{9.45}$$

reproduces the $c = 0$ dispersion relation in (9.40) and (9.42). Consequently, the electric field for (k, ω) is given in terms of the electric field for the point of expansion (k_0, ω_0) through an application of the envelope function $q(z, t)$. Specifically,

$$\vec{E}_{k,\omega}(z, t) = q_0 \hat{n} e^{i[kz-\omega t]} = q(z, t)\hat{n} e^{i[k_0 z-\omega_0 t]} = \frac{q(z, t)}{q_0}\vec{E}_{k_0,\omega_0}(z, t) \tag{9.46}$$

where n is a polarization unit vector. Next, we extend the considerations to treat the role of the envelope function in the presence of the nonlinearity.

To begin a discussion of the nonlinear envelope function it is helpful to make a change of variables in (9.44). Specifically, we change from the variables (z, t) to new variables (z, τ) where $\tau = t - k'z$. Applying this transformation of variables to (9.44), in terms of (z, τ) the envelope equation becomes of the form [53–59]

$$i\frac{\partial q}{\partial z} - \frac{1}{2}k''\frac{\partial^2 q}{\partial \tau^2} + c \mid q \mid^2 q = 0. \tag{9.47}$$

Equation (9.47) is an equation of mathematical physics which occurs in a wide variety of nonlinear problems. It is known as the nonlinear Schrödinger equation because for $c = 0$ it reduces to the general form of the linear Schrödinger equation. An important solution of (9.44) and (9.47) which only exists in the presence of the nonlinearity is that of a bright soliton. This is a pulse solution given by the form [53–58]

$$q(z, \tau) = \pm A e^{ibz}\text{sech}(a\tau) \tag{9.48a}$$

where

$$A^2 = -\frac{k''a^2}{c} \tag{9.48b}$$

and

$$a^2 = -\frac{2b}{k''}. \tag{9.48c}$$

Notice in this solution that $\tau = t - k'z$ so that the pulse translates along the z-axis with a velocity $\frac{1}{k'}$. In addition to the bright soliton other types of solitons also exist in the solutions of the nonlinear Schrödinger equation. These are not discussed here.

9.5.3.3 Parametric amplifier and oscillator theory for optical fibers

The result in (9.44) can be generalized to treat second harmonics, parametric amplification, and parametric oscillation. A simplified treatment giving the gist of the theory is made by replacing the Kerr nonlinearity by an appropriate nonlinearity for the parametric effects. This development follows along the same path as our earlier discussions leading to (9.27).

In this regard, consider the $c = 0$ modal wave functions of the fiber at the three different frequency ω_1, ω_2, ω_3 where $\omega_3 = \omega_1 + \omega_2$. The differential envelope equations for these wave functions are obtained from the dispersion relation expansions about $k_{1,0}$, $k_{2,0}$, $k_{3,0}$ and $\omega_{1,0}$, $\omega_{2,0}$, $\omega_{3,0}$.

These dispersion relations are given by

$$k_j - k_{j,0} = k'(\omega_{0,j})(\omega_j - \omega_{0,j}) + \frac{1}{2}k''(\omega_{0,j})(\omega_j - \omega_{0,j})^2 \quad \text{for} \quad = 1, 2, 3, \tag{9.49}$$

and with the identifications in (9.43) yield the system of three independent differential equations

$$i\left(\frac{\partial q_{\omega_j}}{\partial z} + k_j'\frac{\partial q_{\omega_j}}{\partial t}\right) - \frac{1}{2}k_j''\frac{\partial^2 q_{\omega_j}}{\partial t^2} = 0 \text{ for } = 1, 2, 3. \tag{9.50}$$

The solutions of (9.50) represent three noninteracting guided waves propagating down the fiber.

As in (9.45) the plane wave solutions of the set in (9.50) are of the general forms [53–59]

$$q_{\omega_1}(z, t) = \frac{1}{2}\left[q_1(z)e^{i\left[(k_1-k_{1,0})z-(\omega_1-\omega_{1,0})t\right]} + c.\,c\right] \tag{9.51a}$$

$$q_{\omega_2}(z, t) = \frac{1}{2}\left[q_2(z)e^{i\left[(k_2-k_{2,0})z-(\omega_2-\omega_{2,0})t\right]} + c.\,c\right] \tag{9.51b}$$

$$q_{\omega_3}(z, t) = \frac{1}{2}\left[q_3(z)e^{i\left[(k_3-k_{3,0})z-(\omega_3-\omega_{3,0})t\right]} + c.\,c\right] \tag{9.51c}$$

with each, respectively, corresponding to the frequencies ω_1, ω_2, ω_3. (Notice that for the $c = 0$ noninteracting system in (9.45) the $q_1(z) = 2q_0$ are constants independent of z.) In the following we shall again be interested, as in the treatment leading to (9.23), in studying the different plane wave components of (9.51).

The nonlinearity is now introduced into (9.50) by adding to them the appropriate frequency components of a nonlinear field coupling term. This term is specifically designed to introduce the leading order nonlinear coupling expected from the interaction of the fields with the medium. The introduction is done in a manner which is roughly similar to our discussions leading to (9.27) where the time derivative of (9.15) was introduced as the nonlinearity. Specifically, the form of (9.15) and its time derivative in (9.22) are now generalized to represent a nonlinear response of the medium to the fields in terms of a set of functions

$$S_j(z,t) = e_j\left[q_{\omega_1}(z,t),\, q_{\omega_2}(z,t),\, q_{\omega_3}(x,t)\right] \tag{9.52}$$

for $j = 1, 2, 3$ representing the generated field at ω_j. These terms arise from the nonlinearity of the medium and the fields at the other frequencies, and their various accommodated phase factors. The frequency components of this term are next introduced on the left side of (9.50) to model the response of the nonlinear medium to its applied fields.

In this way, it follows from (9.50) that for $j = 1, 2, 3$

$$\left[i\left(\frac{\partial}{\partial z} + k_j'\frac{\partial}{\partial t}\right) - \frac{1}{2}k_j''\frac{\partial^2}{\partial t^2}\right]\left(q_j e^{i\left[(k_j - k_{j,0})z - (\omega_j - \omega_{j,0})t\right]}\right) + S_j = 0 \tag{9.53a}$$

where, for the case in which $\omega_3 = \omega_1 + \omega_2$

$$S_1(z,t) = e_1\left[q_{\omega_1}(z,t),\, q_{\omega_2}(z,t),\, q_{\omega_3}(x,t)\right], \tag{9.53b}$$

$$S_2(z,t) = e_2\left[q_{\omega_1}(z,t),\, q_{\omega_2}(z,t),\, q_{\omega_3}(x,t)\right], \tag{9.53c}$$

$$S_3(z,t) = e_3\left[q_{\omega_1}(z,t),\, q_{\omega_2}(z,t),\, q_{\omega_3}(x,t)\right]. \tag{9.53d}$$

Similarly, terms representing dissipative losses can also be added into (9.53) so that the resulting system is made to mimic the parametric phenomena discussed earlier.

The formulation in (9.53) can be now be used as a basic model of the parametric properties of optical fibers. This proceeds as in the earlier discussion given in (9.27) through (9.37) and is not pursued here.

9.5.4 Examples of systems: photonic crystal slabs

Second harmonic generation and parametric properties have also been investigated in the context of photonic crystal slabs [60, 61]. In this type of system, a photonic crystal pattern, which is tailored to overcome phase matching problems, is incorporated into a dielectric slab. The pattern is manipulated to optimize the nonlinear effects and the efficiency of the generation of the frequencies of light that are of interest. For the particular studies referenced here, the focus has either been on experimental results [60] or computer simulation studies [60, 61] designed to optimize the nonlinear effects. The reader is referred to the literature for a complete description of these simulation and experimental studies.

9.6 SRRs revisited

As a final point, we remark on some of the structures used in engineering realizations of SRR metamaterials. One of the first structures proposed as an SRR unit is based on the SRR feature presented in figure 4.1 but is a little more complicated. Pendry *et al* [36] initially proposed an SRR in which two of the C-shaped structures in figure 4.1 are composed into a single unit in which one large C contains a second small C such that the gaps of the two Cs are 180 degrees in opposition to one another [36, 37]. These were subsequently shown [37–39] to have essentially the same type of properties as the capacitive loaded loop antenna of figure 4.1. In the SRR proposed by Pendry *et al* [36] the magnetic resonance feature of the unit depends on more parameters of the system than in the simple capacitive loaded loop antenna, i.e., the inner and outer radii of the rings, the gaps in the rings, and the gap between the inner and outer Cs. This allows for a system which is easier to adjust to engineering applications. In addition, the double C structure reduces electric dipole effects associated with the capacitive gaps of the Cs and provides for capacitive effects arising between the inner and outer Cs [40]. Soon after the proposal an experimental application to the magneto-inductive waves in one-dimensional chains [18–21, 41, 42] was made [43] as well as in the first designs of negative indexed media [44].

Since Pendry's suggestion of an SRR structure a number of other configurations have been put forth, some of which are specifically tailored to handle different frequency ranges of applications. These include SRR features of: U-shape [45], omega structures [46], S-shape [47], and the single C-shape [48]. The representations of many of these structures in terms of equivalent circuit models have been provided [37–39, 49], expressing their basic *LRC* properties in terms of their specific structural details. The frequency ranges over which the different types of SRRs operate to provide a negative permeability response, however, in all cases are found to be greatly limited. This continues to be a basic problem in the design of negative indexed materials.

References

[1] Taflove A 1995 *Computational Electrodynamics: The Finite-Difference Time-Domain Method* (Boston, MA: Artech House)

[2] Yeh P 2005 *Optical Waves in Layered Media* (New York: Wiley)

[3] Boyd R W 2003 *Nonlinear Optics* 2nd edn (Amsterdam: Academic)

[4] Mills D L 1998 *Nonlinear Optics* (Berlin: Springer)

[5] Chen W and Mills D L 1987 Optical response of a nonlinear dielectric film *Phys. Rev.* B **35** 524

[6] Chen W and Mills D L 1987 Optical response of nonlinear multilayer structure: bilayers and superlattices *Phys. Rev.* B **36** 6269

[7] Chen W and Mills D L 1987 Gap solitons and the nonlinear optical response of super lattices *Phys. Rev. Lett.* **58** 160

[8] Agranovich V M, Kiselev S A and Mills D L 1991 Optical multistability in nonlinear superlattices with very thin layers *Phys. Rev.* B **44** 10917

[9] McGurn A R 2005 Impurity mode techniques applied to the study of light sources *J. Phys. D: Appl. Phys.* **38** 2338

[10] Marder M P 2010 *Condens. Matter Phys.* 2nd edn (Hoboken, NJ: Wiley)

[11] Yacomotti A M *et al* 2006 All-optical bistable band-edge modes in a two-dimensional photonic crystal *Appl. Phys. Lett.* **88** 231107

[12] Lyubchanskii I, Dadoenkova N N, Zabolotin A E, Lee Y and Rasing T 2008 Optical bistability in one-dimensional magnetic photonic crystal with two defect layers *J. Appl. Phys.* **103** 07B321

[13] McGurn A R and Maradudin A A 1994 Localization of light in linear and nonlinear random layered dielectric systems that are periodic on average *Proc. SPIE* **2117** 12

[14] Scalora M, Dowling J P, Bowden C M and Bloemer M J 1994 Optical limiting and switching of ultrashort pulses in nonlinear photonic band gap materials *Phys. Rev. Lett.* **73** 1368

[15] Scalora M, Dowling J P, Bowden C M and Bloemer M J 1994 The photonic band edge optical diode *J. Appl. Phys.* **76** 2023

[16] Kivshar Y S and Agrawal G P 2003 *Optical Solitons* (Amsterdam: Academic)

[17] Slusher R E and Eggleton B J 2013 *Nonlinear Photonic Crystals* (Berlin: Springer)

[18] Eleftheriou M, Lazarides N and Tsironis G P 2008 Magnetoinductive breathers in metamaterials *Phys. Rev.* E **77** 036608

[19] Eleftheriades G V 2009 EM transmission-line metamaterials *Mater. Today* **12** 30

[20] Kourakis I, Lazarides N and Tsironis G P 2007 Self-focusing and envelope pulse generation in nonlinear magnetic metamaterials *Phys. Rev.* E **75** 067601

[21] Lapine M, Gorkunov M and Ringhofer K H 2003 Nonlinearity of a metamaterial arising from diode inserts into resonant conductive elements *Phys. Rev.* E **67** 065601

[22] McGurn A R 2007 Nonlinear optical media in photonic crystal circuits: device applications *Nonlinear Phenomena Research Perspectives* ed C W Wang (Hauppauge: Nova Science) ch 3 pp 329–67

[23] Yanik M F, Fan S, Soljacic M and Joannopoulos J D 2003 All-optical transistor action with bistable switching in a photonic crystal cross-waveguide geometry *Opt. Lett.* **28** 2506

[24] Soljacac M, Johnson S G, Fan S, Ibanescu M, Ippen E and Joannopoulos J D 2002 Photonic-crystal slow-light enhancement of nonlinear phase sensitivity *J. Opt. Soc. Am.* **B 19** 2052

[25] Soljacic M, Ibanescu M, Johnson S G, Fink Y and Joannopoulos J D 2002 Optimal Bistable switching in nonlinear photonic crystals *Phys. Rev.* E **66** 055601

[26] Yariv A 1971 *Introduction to Optical Electronics* (New York: Holt, Rinehart, and Winston)

[27] Band Y B 2006 *Light and Matter: Electromagnetism, Optics, Spectroscopy, and Lasers* (Chichester: Wiley)

[28] Peng L-P, Hsu C-C and Shih Y C 2003 Second-harmonic Green generation from two-dimensional nonlinear photonic crystal with orthorhombic lattice structure *Appl. Phys. Lett.* **83** 3447

[29] Waser R (ed) 2005 *Nanoelectronics and Information Technology: Advanced Electronic Materials and Novel Devices* 2nd edn (Weinheim: Wiley)

[30] Cohen B E 2010 Biological imaging: beyond fluorescence *Nature* **467** 407

[31] Shen Y R 1989 Surface properties probed by second-harmonic and sum-frequency generation *Nature* **337** 519
[32] Lee J L, Tymchenko M, Argyropoulos C, Chen P-Y, Lu F, Demmerie F, Boehm G, Amann M-C, Alu A and Belkin M A 2014 Giant nonlinear response from plasmonic metasurfaces coupled to intersubband transitions *Nature* **511** 65
[33] Matroell J and Corbalan R 1994 Enhancement of second harmonic generation in a periodic structure with a defect *Opt. Commun.* **108** 319
[34] Scalora M, Bloemer M J, Manka A S, Dowling J P, Bowden C M, Viswanathan R and Haus J W 1997 Pulsed second-harmonic generation in nonlinear, one-dimensional, periodic structures *Phys. Rev.* A **56** 3166
[35] D'Aguanno G *et al* 2001 Photonic band edge effects in finite structures and applications to $\chi^{(2)}$ interactions *Phys. Rev.* E **64** 016609
[36] Pendry J B, Holden A J, Robbins D J and Stewart W J 1999 Magnetism from conductors and enhanced nonlinear phenomena *IEEE Trans. Microw. Theory Tech.* **47** 2075
[37] Engheta N and Ziolkowski R W (ed) 2006 *Metamaterials: Physics and Engineering Explorations* (New York: Wiley-IEEE)
[38] Hrabar S and Bartolic J 2002 Capacitively loaded loop as basic element of negative permeability metamaterial *Proc. European Microwave Conf. (Milan)* vol 2 357
[39] Hrabar S and Bartolic J 2002 Simplified analysis of split ring resonator used in backward meta-materials *Proc. Int. Conf. Mathematical Methods in Electromagnetic Theory (Kiev)* vol 2 500
[40] Cai W and Shalaev V 2001 *Optical Metamaterials: Fundamental and Applications* (New York: Springer)
[41] Giri P, Choudhary K, Sen Gupta A, Bandyopadhyay A K and McGurn A R 2011 Klein–Gordon equation approach to nonlinear split-ring resonator based metamaterials: one-dimensional systems *Phys. Rev.* B **84** 155429
[42] Shamonina E, Kalinin V A, Ringhofer K H and Solymar L 2002 Magnetoinductive waves in one, two, and three dimensions *J. Appl. Phys.* **92** 6252
[43] Shadrivov I V, Reznik A N and Kivshar Yu S 2007 Magnetoinductive waves in arrays of split-ring resonators *Physica* B **394** 180
[44] Smith D R, Padilla W J, Vier D C, Nemat-Nasser S C and Schultz S 2000 Composite medium with simultaneously negative permeability and permittivity *Phys. Rev. Lett.* **84** 4184
[45] Enkrich C, Wegener M, Linden S, Burger S, Zschierich L, Schmidt F, Zhou J F, Koschny T and Soukoulis C M 2005 Magnetic metamaterials at telecommunication and visible frequencies *Phys. Rev. Lett.* **95** 203901
[46] Braaten B D, Scheeler R P, Reich M, Nelson R M, Bauer-Reich C, Glower J and Owen G J 2010 Compact metamaterial-based UHF RFID antennas: deformed omega and split-ring resonator structures *ACES J.* **25** 530
[47] Chen H S, Ran L X, Huangfu J T, Zhang X M, Chen K S, Grzegorczyk T M and Kong J A 2005 Magnetic properties of s-shaped split-ring resonators *Prog. Electromagn. Res.* **51** 231
[48] Yen T J and Lai Y-C 2011 A plasmonic biosensor demonstrates high sensitivity and long-distance detection *SPIE Newsroom* August 10
[49] Baena J D, Bonache J, Martin F, Sillero R M, Lopetegi T, Laso M A G, Gil I, Portillo M F and Sorolla M 2005 Equivalent-circuit models for split-ring resonator and complementary split-ring resonators coupled to planar transmission lines *IEEE Trans. Microw. Theory Tech.* **53** 1451

[50] Liu X, Yang Z and Cao H 2016 Molecular structures and second-order nonlinear optical properties of ionic organic crystal materials *Crystals* **6** 158
[51] Cole J M 2003 Organic materials for second-harmonic generation: advances in relating structure to function *Phil Trans R Soc Lond* **361** 2751
[52] Klein M W 2007 *Nonlinear optics of metallic photonic crystal slabs and planar metamaterials PhD Thesis University Karlsruhe Mainz*
[53] Mollenauer L F and Gordon 2006 *Solitons in Optical Fibers: Fundamentals and Applications* (Amsterdam: Elsevier Science & Technology)
[54] Hasegawa A and Tappert F 1973 Transmission of stationary nonlinear optical pulse in dispersive dielectric fibers. I. Anomalous dispersion *Appl. Phys. Lett.* **23** 142
[55] Hasegawa A and Tappert F 1973 Transmission of stationary nonlinear optical pulse in dispersive dielectric fibers. I. Normal dispersion *Appl. Phys. Lett.* **23** 171
[56] Mitchke F 2016 *Fiber Optics* (Berlin: Springer)
[57] Jackson J D 1999 *Classical Electrodynamics* 3rd edn (New York: Willey)
[58] McGurn A R 2020 *Introduction to Photonic Crystals and Metamaterials* (Santa Rosa, CA: Morgan & Claypool)
[59] Zhang L 2016 *Ultra-Broadly Tunable Light Sources Based on Nonlinear Effects in Photonic Crystal Fibers* (New York: Springer)
[60] Cristan N 2010 Second harmonic generation in sub-diffractive two-dimensional photonic crystals *PhD Thesis* Universitat Politecnica de Catalunya
[61] Conti C, Di Falco A and Assanto G 2004 Parametric oscillation in photonic crystal slabs 3-D time-domain analysis *IEEE Photonics Technol. Lett.* **4** 1041

IOP Publishing

Introduction to Nonlinear Optics of Photonic Crystals and Metamaterials (Second Edition)

Arthur R McGurn

Chapter 10

Disordered systems: site impurities and random media

In this chapter a discussion is given of some of the basic properties of optical systems in the presence of disorder [1–12]. The emphasis is on introducing a variety of general techniques that have been developed to treat disorder and in forming a qualitative understanding of the changes in the properties of materials containing impurities. Many of the techniques presented here are common to the studies of optical, acoustic, and electron systems, and this interconnection is addressed in parts of the chapter. The interrelation and similarity of the mathematics representing these different types of materials is pointed out in the course of the treatment and in the review of recent results in these fields. In this regard, phenomena present in optical systems are expected to have counterparts in the study of acoustics and electronics. In addition to these technical points, a qualitative understanding of the nature of some of the phenomena observed in disordered media is a focus, with a review given of recent work that has been done associated with photonic crystals and metamaterials [9]. This should aid the student to progress quickly in initiating original studies of their own in these new areas of physics and technology.

The treatment begins with a discussion of single impurity sites composed of both linear and nonlinear media [1, 3–6]. In photonic crystals these form sites about which bound states occur as isolated energy states within stop bands and/or as resonant modes developed within the pass bands. Such modes have been useful in the design of resonant cavities, in the uploading or downloading of radiation from waveguide channels, and in the manipulation of radiation lifetimes in respect to transitions found in a variety of optical systems [9]. In this regard, the waveguides of interest to us here are formed in the photonic crystal as a line of site impurities which are adjusted to bind guided modes to them at frequencies in the stop bands of the photonic crystal. Some of the work done on example systems and for recent applications is summarized.

In addition, a focus is on finite clusters of site impurities and how the symmetry properties of a cluster of sites show up in the number and discrete frequencies of bound localized modes in the system [3–6]. This introduces the ideas of group theory into the discussions as a method for the design and development of bound state modes of different characteristic symmetries relative to the photonic crystal lattice. A summary of developments directed towards an understanding of impurity clusters and their bound state and resonant modes is then given for both photonic crystals and metamaterial designs [6–12].

The interaction of waveguide modes with off-channel sites and other waveguides is also a topic. This is a focus of interest in the development of optical circuits [9].

Following the treatment of sites and clusters of sites, an expansion of the discussions to systems in the presence of general volume and surface disorder is made. A particular point of interest in the discussions is on an introduction to the properties of Anderson localization in optically disordered media [8, 12].

Anderson localization ultimately involves a study of the transport and scattering properties of the excitations in a disorder medium [1, 2, 7–12]. These properties arise from the nature of the wave functions created by the type and intensity of the disorder in the system. If the wave functions in the disordered medium are extended modes covering the entire material, the material exhibits a conductive optical transport. On the other hand if the modes develop into localized states which are confined to a finite subregion of the material, at zero temperature the materials are non-conductive of optical energy through the system. Consequently, the geometry of the wave functions is a key factor entering into the transition from an optical conductor to an optical nonconductor. The question for these systems is, then, are their wave functions extended throughout the medium or are they limited to a finite subregion of the medium? These geometric aspects of the wave functions also are seen in an enhancement in the backscattering from the optical medium, and this effect will form an important part of the discussions.

Whether a disordered material is conductive or not is a question of strong Anderson localization of the wave functions in the material [8, 12]. That is, are the wave functions confined in their spatial extent? If the disorder in the system is not great enough to cause strong Anderson localization, however, it can still exhibit a tendency towards localization by displaying weak localization effects. Weak localization effects are found as a tendency for the system to have large backscattering effects, i.e., the medium exhibits a large scattering of light opposite the incident direction of the light. Both weak and strong Anderson localization arise solely from the phase coherent scattering of waves from a random medium [1, 8, 12]. Whether or not a medium displays weak or strong localization depends only on the strength of the scattering occurring within the system.

In general, it is found that the localization properties of a system only depend on the parameters characterizing the wave nature of the excitations relative to the scale of the disorder in the system. These properties are independent of whether the system is quantum or classical mechanical in its composition. Later this feature is discussed in the context of the Ioffe–Regel criterion for the determination of the conditions under which a transition of a mode in the system is made from an extended to a localized state [1, 7–12]. Consequently, a variety of properties are found in optical, acoustic, and electron systems which display localization effects. In all of this multiplicity of systems both classical and quantum phenomena have been studied which exhibit localization behaviors.

Considerations of localization in optics are focused on both photonic crystal and metamaterial media and how these materials have aided in the observation of localization [1, 7–12]. Optical systems have been of particular interest in the study of localization as, unlike electron systems, two beams of light are, from an engineering standpoint, noninteracting with one another. Electron systems, on the other hand, have complicated electron–electron interactions which make the observation of localization difficult. For the optical systems discussed here, the effects on localization of both linear and nonlinear optical media are treated. In this regard, the presence of nonlinear optical media are found to introduce interesting optical localization properties not found in linear systems.

Prior to the discussions of localizations in optical systems a brief review of Anderson localization of electrons is given [1, 7–12]. This was the original system studied for Anderson localization, and the simplicity of the model of conduction electrons facilitates an easy introduction to the general topic of localization. In this model the scaling theory ideas of localization are developed and used to classify the nature of localization observed in one-, two-, and three-dimensional media. These ideas are extended to the treatment of optical systems and then illustrated from the cited properties observed in a variety of optical media. In particular, the effects on localization in materials which are periodic on average is presented in a number of different photonic crystal studies. These materials are reminiscent of the periodic but disordered systems treated in the electronics of a crystalline medium in the presence of disordering impurities.

As noted earlier, optical materials have of recent formed an important area for the study of localization effects because, unlike the original electronic materials investigated for localization, light does not interact with other light [12]. This is not the case in electronic materials in which two electrons experience a coulomb interaction which can develop into a high degree of correlation between the electrons in the context of, for example, the so-called Fermi liquid theory [1, 2]. The correlations introduced in the Fermi liquid ultimately lead in the extreme limit to the formation of the so-called Wigner solid, being a crystal formed as a lattice of electrons [1, 2]. This should be an extremely self-organized form of electron localization and has nothing to do with disorder. The properties of a highly correlated Fermi liquid are found under certain circumstances to exhibit properties which cannot always be distinguish from Anderson localization effects but arise from electron–electron interactions. In this regard, another type of metal–insulator transition occurring in the absence of disorder and generated solely from the electron–electron interactions is the Mott transition [1, 2]. Light, however, does not develop into such alternative forms of the localized state.

In optical media localization arises solely from the phase coherent scattering in the disorder media. During the course of the discussions, a simple description of the phase coherent scattering effects which contribute to localization is given. These are found to involve scattering paths in the medium which support so-called double passage processes and are present in all Anderson localized electron, optical, or acoustic systems.

An important distinction between optical, acoustic, and electronic systems, however, arises directly from the fundamental difference in the wave equations describing each of these types of materials. While electron systems are characterized by a Schrödinger type of wave equation, the optical and acoustic systems are described by Helmholtz equations [12]. These two classes of equations are fundamentally different forms of wave equations and, consequently, exhibit different types of localization. The difference occurs, specifically, in how randomness enters the equations. In the case of Schrödinger equations, the random disorder enters in a way that helps the systems readily exhibit localization. On the other hand, the Helmholtz equation treats randomness in a way that makes the system less inclined in this regard. This distinction is addressed later in the chapter.

In the following, first single site and clusters of site impurities formed of linear and nonlinear media are discussed, and an introduction to some of the applications of these materials is given. This is followed by a treatment of localization in linear media along with a review of their applications. Finally, the effects of nonlinearity on localized optical systems is presented with a treatment of applications and a description of some of the interesting changes in the localization properties coming from the nonlinearity.

10.1 Sites and clusters and off-channel interactions

In this section the focus is on photonic crystals with a discussion first given of single site and clusters of site impurities formed from linear and nonlinear media which are introduced into photonic crystals [3–6]. This is followed by the treatment of the resonant interaction of off-channel site impurities with the channel modes of a waveguide in a photonic crystal. A final system considered involves the coupling of the modes of an infinite straight waveguide channel to a semi-infinite waveguide where the coupling is mediated by an off-channel site or cluster of off-channel sites. Again, in the photonic crystal waveguide problems both linear and nonlinear media coupling sites are studied.

For these discussions, models are introduced which are based on the difference equation approach developed in chapter 5. These represent waveguides formed by the periodic replacement of photonic crystal cylinders along a crystal axis with cylinders containing impurity media. The waveguide channel formed by such a linear array of replaced sites can be made to transmitted guided modes bound to and propagating along the channel. These guided modes occur at stop band frequencies of the photonic crystal.

In this regard, the formulation of difference equations was earlier used in chapter 8 in studies of impurities within waveguide channels, with a focus on transmission though the channel impurity. The difference equation models present simple analytical treatments to a great variety of optical circuit problems, yielding exact solutions which otherwise require computer simulation approaches for their study. In this chapter the formulation is extended to consider a variety of more complex systems to those treated earlier, some of which offer potential for device applications.

The site impurity problems are important in the design of resonant cavities and in the formulation of features entering into optical circuits [9, 12]. Photonic crystals in particular offer the opportunity to design resonators composed of low loss materials which in general enhance their ability to function as high Q resonators [3]. This enhancement occurs over broader frequency bands than are available from alternative resonator technologies. Single site problems also provide means for the development of resonant couplings between optical circuits. In this respect, the ability to couple different waveguides through linear and nonlinear coupling sites enhances the design capabilities of photonic crystal circuits which depend on the interference of resonantly interacting waves [3, 9]. A variety of different types of methods for uploading and downloading of photonic energy from or to optical components is also provided as well as the shifting of optical energy between different parts of optical circuits.

In the following treatments, the focus is on two-dimensional photonic crystals [3, 9]. These are a common type of system treated in the study of optical circuits. In this respect, many optical circuits that have recently been studied experimentally are printed out in slabs of photonic media, with the periodic patterning developed in the plane of the slab [9]. Nevertheless, in the following the considerations are based on two-dimensional photonic crystals formed of infinite dielectric cylinders. In this geometry, the photonic crystal is periodic in the plane of the photonic crystal and the axes of the dielectric cylinders are perpendicular to the photonic crystal plane. This is an idealization of the slab system which is made to simplify the problems discussed but is found to be a useful and effective model.

10.1.1 Single site and clusters of site impurities in photonic crystals

The formulation we use in the following was introduced in (5.9)–(5.13) of chapter 5 for the impurity modes of a two-dimensional photonic crystal composed with a single site impurity or with an infinite periodic array of dielectric replacement cylinder impurities arrayed along a crystal axis of the photonic crystal [3–6, 9]. It is a difference equation formulation which ultimately comes from a completely general integral equation formula for the scattering from impurities introduced into some of the lattice sites of a two-dimensional photonic crystal. In this regard, however, the added impurity material is assumed to have the same translational symmetry along the direction of the cylinder axes as that of the pure photonic crystal. A brief overview and summary of the theory is presented here, followed by some applications to disorder systems of interest.

For the case in which the electric field is polarized parallel to the cylinder axes the impurity problem involves a solution of the standard integral equation relationship

$$E(\vec{r}, \omega) = \left(\frac{\omega}{c}\right)^2 \int d^2r' G(\vec{r}, \vec{r}'|\, \omega)\delta\varepsilon(\vec{r}')E(\vec{r}', \omega), \tag{10.1}$$

which was discussed earlier in chapter 5 and is also introduced in [3–6, 9]. Here $\delta\varepsilon(\vec{r})$ is the change in the dielectric properties of the photonic crystal arising from the introduction of impurity materials into the photonic crystal, ω is the frequency being investigated, the space vectors are in the plane perpendicular to the cylinder axes, and $G(\vec{r}, \vec{r}'|\, \omega)$ is the Green function of the pure photonic crystal for modes of frequency ω [3, 9]. The result in (10.1) is basically a statement of the scattering problem for scattering from the impurity material within the context of an integral equation format. This replaces the more common scattering formulation in terms of a differential wave equation.

10.1.1.1 Single sites

For the case of a single impurity site in a bulk two-dimensional photonic crystal, $\delta\varepsilon(\vec{r}) = \delta\varepsilon_{00}$ over the connected area of the impurity S in the plane of the periodicity, and $\delta\varepsilon(\vec{r}) = 0$ otherwise. (See figure 10.1(a) for a schematic.) The impurity material is then located only in the region S, and for $\vec{r}' \in S$ it follows from (10.1) that [3–6, 9]

$$E(\vec{r}, \omega) = \left(\frac{\omega}{c}\right)^2 \delta\varepsilon_{00} \int d^2r' G(\vec{r}, \vec{r}'|\, \omega)E(\vec{r}', \omega). \qquad (10.2a)$$

In the case of a small enough impurity such that $\vec{E}(\vec{r}, \omega)$ is approximately constant over the cross sectional area of the impurity in the plane of the photonic crystal S, then $\left(\frac{\omega}{c}\right)^2 \int d^2r' G(\vec{r}, \vec{r}'|\, \omega)E(\vec{r}', \omega) \approx \tilde{\alpha}(\vec{r}, \vec{r}')E(\vec{r}', \omega)$ where $\vec{r}' \in S$ and

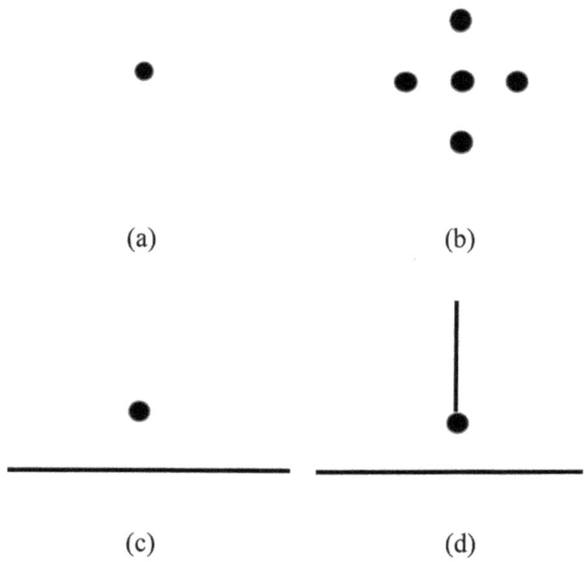

Figure 10.1. Schematic plots of geometries studied for: (a) a single site impurity, (b) a cluster of five impurities, (c) an infinite waveguide channel (represented as a horizontal straight line) interacting with a single off-channel site, and (d) a horizontal infinite waveguide channel (represented by a horizontal line) interacting through a single off-channel site with a vertical semi-infinite waveguide channel (represented by the vertical semi-infinite line). Only the impurity single sites are shown, and the waveguides are represented as lines rather than an array of nearest neighbor sites. The x-axis is horizontal and the y-axis is vertical.

$$\tilde{\alpha}(\vec{r}, \vec{r}') = \left(\frac{\omega}{c}\right)^2 \int_S d^2r' G(\vec{r}, \vec{r}' | \omega) \tag{10.2b}$$

becomes a simple multiplier of the field $E(\vec{r}', \omega)$ which is again constant in $\vec{r}' \in S$. Consequently, under these considerations and taking $\vec{r} \approx \vec{r}' \in S$ in the evaluation of (10.2) the problem reduces to a single linear algebraic equation given by

$$E_0 = g\alpha E_0 \text{ or } 1 = g\alpha \tag{10.3}$$

where $\alpha = \tilde{\alpha}(\vec{r}, \vec{r})/S$, $g = S\delta\varepsilon_{00}$, S is the area of the impurity in the plane of the periodicity, and $E_0 = E(\vec{r}, \omega)$.

It is seen that the solutions of either the general case (10.2), or the special case (10.3), give the real frequencies ω at which bound states exist in the photonic crystal. These must occur in the stop bands of the photonic crystal, and their values depend on the dielectric constant of the impurity material. Also, within the pass bands can be found complex solutions of ω. These represent resonant scattering modes of the system.

In the following we are primarily interested in the bound state modes of the site impurities [3–5]. As a final point, in this regard, it should be noted that the electric field throughout all space can be obtained from (10.2a). This is done by evaluating the integral expression in (10.2a) using the eigenvalue results of (10.3) and considering general $\vec{r}$ rather than just $\vec{r} \approx \vec{r}'$. Such an evaluation then completely solves the bound state problem.

10.1.1.2 Waveguides

Generalizing the earlier discussions to treat a chain of identical small separate impurity sites, each of small cross section S along for example the x-axis of a square lattice photonic crystal, a difference equation treatment follows from similar arguments to those applied to (10.1) and (10.2) for the single site problems [3–6, 9]. (The reader is also referred to our earlier discussions on this system in chapter 5 and to the literature [4, 9] for the details of such a system.) In this manner, considering a photonic crystal formed as an array of parallel axis dielectric cylinders and a waveguide formed along a crystal axis by cylinder replacement, a difference equation for the modes on the chain is generated having the form

$$E_n = g[\alpha E_n + \beta(E_{n-1} + E_{n+1})] \tag{10.4}$$

where the subscripts on the electric fields label the waveguide impurity lattice sites of replacement cylinders where the fields occur. Consequently, E_n is the field on the nth impurity lattice site, and the sites are consecutively label by integers. In addition, here $\beta = \tilde{\alpha}(\vec{r}, \vec{r}')/S$ where $\vec{r}$ and $\vec{r}'$ are in nearest neighbor impurity sites in the waveguide channel and as before in the single site problem α is the onsite coupling. For simplicity the system studied here is limited to onsite and nearest neighbor couplings α and β, respectively. (Note: this limitation is easily relaxed to consider further neighbor interactions.) As before from (10.4) the frequencies of the guided waves in the stop bands of the photonic crystal are directly related to the dielectric constants of the channel sites. These are the modes, bound to and propagating along the channel, of interest for the design of optical circuits.

10.1.1.3 Kerr media

Next consider the introduction of Kerr nonlinear media into the photonic crystal. Upon the introduction of a Kerr nonlinear change in permittivity of the form $\delta\varepsilon(\vec{r}) = \delta\varepsilon_{00}(1 + \lambda\,|E|^2)$, where $|E|^2$ is the modulus square of the field applied to the nonlinear medium, the study of single sites and waveguide channels follows as a direct generalization of the earlier considerations. Specifically, in our treatment small amounts of impurity material are introduced to form single site impurities and/or waveguides. In this way, by arguments similar to those used in arriving at (10.4), it then follows that (10.4) generalizes to [3–6, 9]

$$E_n = g\left[\alpha\left(1 + \lambda\,|E_n|^2\right)E_n + \beta\left(1 + \lambda\,|E_{n+1}|^2\right)E_{n+1} + \beta\left(1 + \lambda\,|E_{n-1}|^2\right)E_{n-1}\right]. \quad (10.5)$$

Notice that now the site couplings are functions of the fields at the site, and for $\lambda = 0$ the result in (10.4) is obtained.

For a single site impurity (10.3) likewise generalizes by the same treatment used to handle a waveguide of Kerr medium impurities. Specifically, such considerations give [3–6, 9]

$$E_0 = g\alpha\left(1 + \lambda\,|E_0|^2\right)E_0 \text{ or } 1 = g\alpha\left(1 + \lambda\,|E_0|^2\right) \quad (10.6)$$

where the impurity site is located at $n = 0$, and unlike the linear case the frequency ω now depends on the electric field.

Both (10.5) and (10.6) now offer solutions for real ω in the stop band, respectively, for waveguide modes and bound site modes. These are framed in terms of the dielectric constants of the impurity materials and the field intensities. Unlike in the problems involving solely linear media, the frequencies at which these solutions occur are found to be dependent on the field amplitudes.

For further details of the difference equation approach the reader may consult chapter 5 and the literature [3–6, 9]. The discussions to follow use the difference equation formulations as a pedagogical tool for developing an understanding of the photonic crystal site bound states and the modes of certain interacting waveguide circuits. A semi-quantitative understand of a variety of optical circuits is then developed based on these simple analytical methods.

10.1.1.4 Single site impurities

The simplest impurity features to treat are single site impurities characterized by (10.3) and (10.6). (See figure 10.1(a) for a schematic representation of a single site impurity within a photonic crystal.) First, we begin with a consideration of the bound state energy solutions of linear dielectric impurities. From these linear media solutions results for the nonlinear system are then generated as their perturbation extensions.

From (10.3) it then follows that [3, 4]

$$\delta\varepsilon_{00} = \frac{1}{\alpha S}, \quad (10.7)$$

where $\alpha = \tilde{\alpha}(\vec{r}, \vec{r})/S$, and the change in dielectric constant $\delta\varepsilon_{00}$ is found to be related to the Green function averaged over the area of the impurity site. Consequently, a knowledge of the functional form of the Green function in ω allows for the complete determination of the bound state frequency ω and wavefunction associated with the impurity $\delta\varepsilon_{00}$ introduced into the crystal. An alternative way of viewing the result, however, is that if one is given ω, a computation of the Green function at this frequency gives a determination of the value of $\delta\varepsilon_{00}$ needed to support a bound state mode at the given frequency ω.

Similarly, considering the case of the Kerr nonlinear site impurity, from (10.6) it is seen that a field dependence is introduced into the denominator of (10.7). It then follows that [3, 4]

$$\delta\varepsilon_{00} = \frac{1}{\alpha\left(1 + \lambda\, |E_0|^2\right)S}. \tag{10.8}$$

The field dependence now enters into the determination of either ω or $\delta\varepsilon_{00}$ as an additional factor. In this respect, it is important to note that for most systems treated in current technologies, it is often the case that $|\lambda||E_0|^2 \ll 1$.

To see the field effects on the bound states for Kerr nonlinear media, consider a first order perturbation treatment of (10.8) in the $|\lambda||E_0|^2 \ll 1$ limit, and its relation to the solutions of (10.7). Notice in the following discussion, that in the limit $\lambda \to 0$ (10.8) reduces to (10.7) and the $|\lambda||E_0|^2 \ll 1$ term generates only small changes in this limit.

If, in the linear medium limit of the impurity system, a site impurity of dielectric change, $\delta\varepsilon_{00}$, is introduced into the system and results in a bound state at frequency ω_0, it follows from (10.3) that [3, 4]

$$1 = g\alpha(\omega_0). \tag{10.9}$$

Consequently, considering (10.6) for the same system upon the introduction of Kerr nonlinearity (i.e., $\lambda \neq 0$), then gives [4]

$$1 = g\alpha(\omega)\left(1 + \lambda\, |E_0|^2\right) = g\alpha(\omega_0 + \Delta\omega)\left(1 + \lambda\, |E_0|^2\right) \tag{10.10}$$

where ω is the new bound state frequency, $\Delta\omega = \omega - \omega_0$, and the Kerr parameter λ has been assumed to be independent of the frequency. Upon expanding $\alpha(\omega_0 + \Delta\omega)$ in $|\Delta\omega| \ll 1$, it then follows that a Taylor expansion about ω_0 to order $\Delta\omega$ yields

$$\alpha(\omega_0 + \Delta\omega) \approx \alpha(\omega_0) + \Delta\omega\alpha'(\omega_0), \tag{10.11}$$

and substituting the result in (10.11) into (10.10) gives the bound state frequency [4]

$$\omega \approx \omega_0 - \frac{\lambda\, |E_0|^2}{g\alpha'(\omega_0)}. \tag{10.12}$$

Notice that the field correction in (10.12) for the bound state frequency in the presence of nonlinearity depends on the sign of $\lambda\, |E_0|^2$, and the correction enters quadratically in the field amplitude. For $\lambda = 0$ in (10.12) the linear limit is again retrieved.

The solutions of the single site impurity problem are observed to be highly symmetric in the lattice, i.e., s-wave bound states prevail. Next, we shall consider clusters of sites. These allow for the introduction of bound states in the stop bands which are less symmetric. Such types of multiple impurities often display multiple bound states, exhibiting a variety of energy levels corresponding to bound state wave functions of different spatial symmetry relative to the crystal lattice. How then does the single site problem generalize to the treatment of clusters of sites?

10.1.1.5 Clusters of site impurities

The development of the theory for bound states of impurity clusters follows along similar lines to that for the single impurity problem just treated [4]. In generalizing the site problem, we go from a focus on a highly symmetric single site to that of a of much lower symmetry finite cluster of single sites. In this regard, as an important new feature of the treatment it is helpful to introduce into the considerations an element of group theory associated with the symmetry properties displayed by the cluster. This allows for the direct introduction of the symmetry of the impurity geometry into the process of solving for the frequencies and wave functions of the bound state modes. Ultimately, the frequency modes of the bound states are found to be associated with the partition into classes of the symmetry operations of the cluster symmetry group.

As an example of this type of approach to clusters, consider a photonic crystal which is in the form of a square lattice array of infinite dielectric cylinders. The square lattice is oriented in space so that a nearest neighbor subset of lattice points are along the x-axis, and, similarly, a nearest neighbor subset of points are along the y-axis. In this arrangement, then, the x–y plane is defined to be perpendicular to the cylinder axes, and the nearest neighbor separation of the lattice sites is denoted by a. As in the case of the earlier discussions for the single impurity site problem, the electric field in the following treatment is considered to be polarized parallel to the cylinder axes.

For a simple cluster problem, consider that a cluster of five impurity sites is introduced into the square lattice photonic crystal just described. One site impurity of the cluster is located at the origin of the square lattice and four additional site impurities are located at the four nearest neighbor sites of the square lattice. A schematic of the cluster is shown in figure 10.1(b). The site impurities introduced to form the cluster are taken to be identical to the single site impurity considered earlier. Each site of the cluster is small enough so that the field is basically constant over its cross section, and none of the sites of the cluster touch each other.

Specifically, if $\delta\varepsilon_0(\vec{r})$ is the change in the dielectric constant for the impurity site at the origin, the addition sites of the cluster are described by the dielectric changes $\delta\varepsilon_0(\vec{r} - a\hat{i})$, $\delta\varepsilon_0(\vec{r} + a\hat{i})$, $\delta\varepsilon_0(\vec{r} - a\hat{j})$, $\delta\varepsilon_0(\vec{r} + a\hat{j})$. In this regard, a statement of the integral equation problem for the cluster of five sites is obtained by considering (10.1) for the case in which $\delta\varepsilon(\vec{r}) = \delta\varepsilon_m(\vec{r})$ where [4]

$$\delta\varepsilon_m(\vec{r}) = \delta\varepsilon_0(\vec{r}) + \delta\varepsilon_0(\vec{r} - a\hat{i}) + \delta\varepsilon_0(\vec{r} + a\hat{i}) + \delta\varepsilon_0(\vec{r} - a\hat{j}) + \delta\varepsilon_0(\vec{r} + a\hat{j}). \quad (10.13)$$

As in the single site consideration, $\delta\varepsilon_0(\vec{r})$ is of small enough cross sectional area that the approximation of (10.2) can be applied and the impurity materials in the five sites summed over in (10.13) are isolated from one another. Consequently, $\delta\varepsilon_0(\vec{r})$ can be taken over directly as the dielectric change used in our earlier considerations of the single site problem.

With regard to $\delta\varepsilon_m(\vec{r})$, the integral equation formulation of the cluster problem under consideration is then given by (10.1). Consequently, upon a substitution, it follows that the electric field is a solution of the integral eigenvalue problem of the form [4]

$$E(\vec{r},\,\omega) = \left(\frac{\omega}{c}\right)^2 \int \mathrm{d}^2r' G(\vec{r},\,\vec{r}'|\,\omega)\delta\varepsilon_m(\vec{r}')E(\vec{r}',\,\omega). \tag{10.14}$$

Here the eigenfunction $E(\vec{r}',\,\omega)$ is the component of the electric field parallel to the cylinder axes, ω is the corresponding modal eigenfrequency, and $G(\vec{r},\,\vec{r}'|\,\omega)$ is the Green's function of the square lattice photonic crystal in the absence of the impurity. Written in terms of (10.13) it is then found that

$$\begin{aligned} E(\vec{r},\,\omega) = \left(\frac{\omega}{c}\right)^2 \int \mathrm{d}^2r' G(\vec{r},\,\vec{r}'|\,\omega)[\delta\varepsilon_0(\vec{r}') + \delta\varepsilon_0(\vec{r}' - a\hat{i}) \\ + \delta\varepsilon_0(\vec{r}' + a\hat{i}) + \delta\varepsilon_0(\vec{r}' - a\hat{j}) + \delta\varepsilon_0(\vec{r}' + a\hat{j})]E(\vec{r}',\,\omega) \end{aligned} \tag{10.15}$$

which is re-expressed as a sum over terms of a similar form given by [5]

$$E(\vec{r},\,\omega) = \left(\frac{\omega}{c}\right)^2 \sum_{i=0}^{4} \int \mathrm{d}^2r' G(\vec{r},\,\vec{r}'|\,\omega)\delta\varepsilon_0(\vec{r}' - \vec{r}_i)E(\vec{r}',\,\omega) \tag{10.16}$$

where $\vec{r}_0 = 0$, $\vec{r}_1 = a\hat{i}$, $\vec{r}_2 = -a\hat{i}$, $\vec{r}_3 = a\hat{j}$, and $\vec{r}_4 = -a\hat{j}$.

Proceeding with the evaluation of the integrals in (10.15) and (10.16) we apply the same methods as those used in the evaluation of the single site problem of (10.2). From this it follows that the sum of integrals on the right-hand side of (10.16) are again reduced to a set of difference equation sums for the fields at the five sites $\vec{r}_0$, $\vec{r}_1$, $\vec{r}_2$, $\vec{r}_3$, and $\vec{r}_4$. The resulting linear equations are now linear matrix equations for the eigenvalue ω corresponding to the eigenvector fields $E_0 = E(\vec{r}_0,\,\omega)$, $E_1 = E(\vec{r}_1,\,\omega)$, $E_2 = E(\vec{r}_2,\,\omega)$, $E_3 = E(\vec{r}_3,\,\omega)$, and $E_4 = E(\vec{r}_4,\,\omega)$ at the sites $\vec{r}_0$, $\vec{r}_1$, $\vec{r}_2$, $\vec{r}_3$, and $\vec{r}_4$ which are generated by evaluating the field on the left-hand side of (10.16) at the sites $\vec{r}_0$, $\vec{r}_1$, $\vec{r}_2$, $\vec{r}_3$, and $\vec{r}_4$ [4].

In this way it follows that the integrals of the Green's functions in (10.16) define a 5×5 matrix of the form [4]

$$\overleftrightarrow{G} = \begin{vmatrix} G_{00} & G_{01} & G_{02} & G_{03} & G_{04} \\ G_{10} & G_{11} & G_{12} & G_{13} & G_{14} \\ G_{20} & G_{21} & G_{22} & G_{23} & G_{24} \\ G_{30} & G_{31} & G_{32} & G_{22} & G_{34} \\ G_{40} & G_{41} & G_{42} & G_{43} & G_{44} \end{vmatrix} = \begin{vmatrix} G_{00} & G_{01} & G_{01} & G_{01} & G_{01} \\ G_{01} & G_{00} & G_{12} & G_{13} & G_{13} \\ G_{01} & G_{12} & G_{00} & G_{13} & G_{13} \\ G_{01} & G_{13} & G_{13} & G_{00} & G_{12} \\ G_{01} & G_{13} & G_{13} & G_{12} & G_{00} \end{vmatrix} \tag{10.17}$$

where the subscripts on the matrix elements indicate the $\vec{r}_0$, $\vec{r}_1$, $\vec{r}_2$, $\vec{r}_3$, and $\vec{r}_4$ sites between which they are defined. Consequently, for example [4]

$$G_{00} = \left(\frac{\omega}{c}\right)^2 \frac{1}{S} \int_S \mathrm{d}^2 r' G(\vec{r}, \vec{r}' | \omega) \text{ for } \vec{r} \approx \vec{r}' \in S, \quad (10.18a)$$

$$G_{01} = \left(\frac{\omega}{c}\right)^2 \frac{1}{S} \int_S \mathrm{d}^2 r' G(\vec{r}, \vec{r}' | \omega) \text{ for } \vec{r} \approx \vec{r}' - \vec{r}_1 \in S, \quad (10.18b)$$

$$G_{12} = \left(\frac{\omega}{c}\right)^2 \frac{1}{S} \int_S \mathrm{d}^2 r' G(\vec{r}, \vec{r}' | \omega) \text{ for } \vec{r} \approx \vec{r}' - 2\vec{r}_1 \in S, \quad (10.18c)$$

$$G_{13} = \left(\frac{\omega}{c}\right)^2 \frac{1}{S} \int_S \mathrm{d}^2 r' G(\vec{r}, \vec{r}' | \omega) \text{ for } \vec{r} \approx \vec{r}' - \vec{r}_1 - \vec{r}_3 \in S. \quad (10.18d)$$

Here S is defined in the x–y plane as the small element of impurity area surrounding the position $\vec{r}_0$ of the center site of the impurity cluster which is located at the origin. Specifically, it is the region over which the term G_{00} is defined. The other integrals in (10.18b) through (10.18d), however, involve relative shifts between $\vec{r}$ and $\vec{r}'$ involving one or two lattice constants along the x- and/or y- directions. Also, note that in arriving at the form of the matrix elements on the far right-hand side of the equation in (10.17), the symmetries of the square lattice of the unperturbed Green's function have been applied. This asserts the equality of some of the matrix elements which are related to one another by lattice symmetries. Consequently, for example $G_{00} = G_{11} = G_{22} = G_{33} = G_{44}$, $G_{01} = G_{02} = G_{03} = G_{04}$, etc. through the symmetry group of the square lattice.

Within our approximations the matrix in equation (10.18) now forms the relationship between the fields on the left and right sides of the integral equation in equation (10.16). This relationship is next written out.

Writing the electric field vectors on either of the two sides of the integral equation in (10.16) in the general form [4]

$$\mathbb{E} = \begin{vmatrix} E(\vec{r}_0, \omega) \\ E(\vec{r}_1, \omega) \\ E(\vec{r}_2, \omega) \\ E(\vec{r}_3, \omega) \\ E(\vec{r}_4, \omega) \end{vmatrix} = \begin{vmatrix} E_0 \\ E_1 \\ E_2 \\ E_3 \\ E_4 \end{vmatrix} \quad (10.19)$$

the integral equation in (10.16) is found to be reduced to the algebraic eigenvalue problem

$$\begin{vmatrix} E_0 \\ E_1 \\ E_2 \\ E_3 \\ E_4 \end{vmatrix} = g \begin{vmatrix} G_{00} & G_{01} & G_{01} & G_{01} & G_{01} \\ G_{01} & G_{00} & G_{12} & G_{13} & G_{13} \\ G_{01} & G_{12} & G_{00} & G_{13} & G_{13} \\ G_{01} & G_{13} & G_{13} & G_{00} & G_{12} \\ G_{01} & G_{13} & G_{13} & G_{12} & G_{00} \end{vmatrix} \begin{vmatrix} E_0 \\ E_1 \\ E_2 \\ E_3 \\ E_4 \end{vmatrix} \quad (10.20a)$$

so that

$$\mathbb{E} = g\overleftrightarrow{G}\mathbb{E}. \tag{10.20b}$$

Here $g = S\delta\varepsilon_{00}$ and $\delta\varepsilon_{00}$ is the constant value of $\delta\varepsilon_0(\vec{r})$ over the region in which it is nonzero.

From (10.20) the problem of the bound state modes of the five impurity sites is seen to involve the determination of the real eigenvalues ω and their associated eigenvectors $\mathbb{E}(\omega)$ of a simple matrix equation. However, by first making an application of some basic ideas from group theory to the form of the impurity problem in (10.20) it is possible to arrive at a very nice exposition of the nature of the modes within the impurity system. This is now addressed.

10.1.1.6 Group theory reduction

Considering the symmetries of the impurity composed of five sites [4], it is found that the group formed by these operations for the cluster is the same as the symmetry group of a square. Specifically, there are eight elements in the symmetry group of the square. These include: rotations about the center site of the cluster by 90°, 180°, 270°; reflections in the horizontal *x*- and vertical *y*-axes; reflections in the two diagonals of the square; and the identity element which leaves the square unchanged. The total set of these operations generate a complete, closed, group multiplication table under all of their compositions.

The eight symmetry elements comprising the group can be partitioned into classes, each of which is composed of similar operations [9]. In the case of the symmetry group of the square, this partition gives five classes which include: 1. The identity which does nothing to the square; 2. The rotation by 180° about an axis perpendicular to the *x*–*y* plane and passing through the center site; 3. The two rotations by 90° and 270° about an axis perpendicular to the *x*–*y* plane and passing through the center site; 4. The reflections about the horizontal *x*- and vertical *y*-axes; 5. The reflections in the two diagonals of the square. Each of these classes contain symmetry operations which are the same in nature but applied to the square in a different way [9]. For example, the reflections about the horizontal *x*- and vertical *y*-axes basically do the same thing to the square about different axes which are similarly related to the square geometry. In this format, however, the reflections just cited are different in nature from the reflections in the two diagonals of the square whose axes are differently related to the square geometry. The partition of the group elements into classes is very useful as it has a fundamental importance in determining the nature of the energy levels displayed by the impurity bound states of the cluster.

In this regard, there is a fundamental result from group theory that now applies to our cluster problem [9]: as a first point it should be noted that the bound state energy levels associated with the impurity sites can exhibit energy level degeneracies, and the possible energy degeneracies of the bound states are related to the number of classes of the symmetry group. More specifically, the number of elements in the group must equal a sum of squared integers, and the number of squared integers in the sum equals the number of classes in the group. For groups of a small number of symmetries this offers a good indicator of the nature of the energy levels found in the bound state spectrum.

From these general group theory considerations [9], the spectrum of the five-site cluster can be determined. Consequently, for the group of symmetry elements of the square there are five classes so that for the eight elements of the symmetry group it follows that [5, 9]

$$8 = 2^2 + 1^2 + 1^2 + 1^2 + 1^2. \tag{10.21}$$

Here it is also important to note that there should be an energy level which is singly degenerate. This corresponds to a wave function which remains the same under each of the symmetry operations of the group. Under this restriction it is found that the sum in (10.21) is a unique representation of 8 in terms of the squares of appropriate positive integers.

These general results from the theory of groups and classes are very helpful in understanding the bound state energies associated with small clusters of sites in impurity systems. Specifically, they allow for the determination of the degeneracies of the bound state energies of the impurity problem. Next the full algebraic approach to finding the energy levels and their eigenfunctions is addressed.

10.1.1.7 Full solution

Another approach to obtaining information about the bound states of the cluster is from a direct solution of (10.20). For this treatment it is useful to first apply a unitary transformation to (10.20). The unitary transformation is chosen to put the matrix $\overleftrightarrow{G}$ into a block diagonal form in which its solution is more manageable. In the present case, the particular transformation used was originally developed to study problems in molecular bonding and magnetic impurities [9]. In this respect, the transformation helps separate the solutions into s-, p-, and d-symmetry types [4].

Specifically, we apply to (10.20) the similarity transformation [4]

$$U^{+}\mathbb{E} = gU^{+}\overleftrightarrow{G}UU^{+}\mathbb{E} \tag{10.22}$$

where the unitary matrix of the transformation is given by

$$U = \begin{vmatrix} 1 & 0 & 0 & 0 & 0 \\ 0 & \frac{1}{2} & \frac{1}{\sqrt{2}} & 0 & \frac{1}{2} \\ 0 & \frac{1}{2} & -\frac{1}{2} & 0 & \frac{1}{2} \\ 0 & \frac{1}{2} & 0 & \frac{1}{\sqrt{2}} & -\frac{1}{2} \\ 0 & \frac{1}{2} & 0 & -\frac{1}{\sqrt{2}} & -\frac{1}{2} \end{vmatrix}, \tag{10.23}$$

with $U^+U = UU^+=1$. (Notice that similar transformations for other finite clusters can be taken from group theory studies of molecules and magnetic impurity problems [9].) Evaluating (10.22) and (10.23), it then follows that [4]

$$\mathbb{G} = U^+\overleftrightarrow{G}U = U^+ \begin{vmatrix} G_{00} & G_{01} & G_{01} & G_{01} & G_{01} \\ G_{01} & G_{00} & G_{12} & G_{13} & G_{13} \\ G_{01} & G_{12} & G_{00} & G_{13} & G_{13} \\ G_{01} & G_{13} & G_{13} & G_{00} & G_{12} \\ G_{01} & G_{13} & G_{13} & G_{12} & G_{00} \end{vmatrix} U = \begin{vmatrix} G_{00} & 2G_{01} & 0 & 0 & 0 \\ 2G_{01} & G_s & 0 & 0 & 0 \\ 0 & 0 & G_p & 0 & 0 \\ 0 & 0 & 0 & G_p & 0 \\ 0 & 0 & 0 & 0 & G_d \end{vmatrix} \qquad (10.24)$$

where $G_s = G_{00} + G_{12} + 2G_{13}$, $G_p = G_{00} - G_{12}$, and $G_d = G_{00} + G_{12} - 2G_{13}$. The block diagonal form in equation (10.24) is seen to reduce the complexity of the problem which is now treated analytically. In addition, it should be noted that (10.24) is reduced to s-, p-, and d-symmetry components as indicated by the subscripts on the far right of (10.24).

Similarly, writing the transformed eigenvector as $\tilde{\mathbb{E}} = U^+\mathbb{E}$ it follows that [4]

$$\tilde{\mathbb{E}} = \begin{vmatrix} E_0 \\ E_s \\ E_p \\ E_p \\ E_d \end{vmatrix}, \qquad (10.25)$$

and (10.22) then takes the form

$$\tilde{\mathbb{E}} = g\mathbb{G}\tilde{\mathbb{E}}. \qquad (10.26)$$

Here the original eigenvalue problem in (10.22), involving a 5×5 matrix, is simplified into the form of a 2×2 matrix eigenvalue problem for the s-symmetry modes and three 1×1 matrix eigenvalue problems for the p- and d-symmetry modes. In this matrix reduction it is found from (10.26) that there are three non-degenerate eigenvalues and a single doubly degenerate p-symmetry eigenvalue. Along with these eigenvalues are a correspond set of simply expressed symmetrized eigenfunctions.

Continuing from (10.26) it follows that of the three 1×1 matrix eigenvalue problems, two are of the identical forms

$$E_p = g_p G_p E_p, \qquad (10.27a)$$

and one is given by

$$E_d = g_d G_d E_d. \qquad (10.27b)$$

In addition, the 2×2 matrix eigenvalue problem for the two s-symmetry states is [4]

$$\begin{vmatrix} E_0 \\ E_s \end{vmatrix} = g_s \begin{vmatrix} G_{00} & 2G_{01} \\ 2G_{01} & G_s \end{vmatrix} \begin{vmatrix} E_0 \\ E_s \end{vmatrix}. \qquad (10.28)$$

This remaining 2×2 matrix eigenvalue problem is now solved for its two non-degenerate s-symmetry eigenfunctions. Upon obtaining the solution for the system of equations in (10.27) to (10.28) it will be seen that degenerate states of the system are from the two identical E_p equations while the remaining solutions are all non-degenerate.

Considering (10.28) it follows from a little standard algebra that the eigenvalues of the s-modes reduce to [4]

$$g_{s\pm} = \frac{1}{2}\frac{G_{00} + G_s \pm \sqrt{(G_{00} - G_s)^2 + 16G_{01}^2}}{G_{00}G_s - 4G_{01}^2} \tag{10.29a}$$

where the corresponding eigenvectors $\begin{vmatrix} E_0 \\ E_s \end{vmatrix}$ are of the form

$$\frac{E_0}{E_s} = 2\frac{G_{01}}{G_s - g_{s\pm}[G_{00}G_s - 4G_{01}^2]}. \tag{10.29b}$$

Similarly, for the p- and d-modes it is found from (10.27) that

$$g_p = [G_p]^{-1}, \tag{10.30}$$

$$g_d = [G_d]^{-1}. \tag{10.31}$$

In the expressions of (10.29) through (10.31) the forms of the symmetrized Green functions are defined by [4]

$$G_{00} = \left(\frac{\omega}{c}\right)^2 \sum_r \int \frac{d^2k}{(2\pi)^2} G_r(\vec{k}), \tag{10.32a}$$

$$G_{01} = \left(\frac{\omega}{c}\right)^2 \sum_r \int \frac{d^2k}{(2\pi)^2} \gamma_{\vec{k}} G_r(\vec{k}), \tag{10.32b}$$

$$G_s = \left(\frac{\omega}{c}\right)^2 \sum_r \int \frac{d^2k}{(2\pi)^2} (1 + 2\nu_{\vec{k}} + \gamma_{2\vec{k}}) G_r(\vec{k}), \tag{10.32c}$$

$$G_p = \left(\frac{\omega}{c}\right)^2 \sum_r \int \frac{d^2k}{(2\pi)^2} (1 - \gamma_{2\vec{k}}) G_r(\vec{k}), \tag{10.32d}$$

and

$$G_d = \left(\frac{\omega}{c}\right)^2 \sum_r \int \frac{d^2k}{(2\pi)^2} (1 - 2\nu_{\vec{k}} + \gamma_{2\vec{k}}) G_r(\vec{k}). \tag{10.32e}$$

Here the integrals in (10.32) are evaluated over the first Brillouin zone of the photonic crystal, and in the index sum r labels all the distinct energy bands in the first

Brillouin zone having the same wavevector $\vec{k}$. Notice that the expressions for the Green functions are written in terms of the band dispersion relations $\epsilon_r(\vec{k})$ of the square lattice photonic crystal and the integrands in (10.32) involve the functional form [4]

$$G_r(\vec{k}) = \frac{E_r^* E_r}{\epsilon_r - \frac{\omega^2}{c^2}}. \tag{10.33a}$$

In addition, the integrands include the wavevector functions given by

$$\nu_{\vec{k}} = \cos k_x a \cos k_y a \tag{10.33b}$$

and

$$\gamma_{\vec{k}} = \frac{1}{2}\left[\cos k_x a + \cos k_y a\right] \tag{10.33c}$$

for the wavevectors $\vec{k} = (k_x, k_y)$ in the first Brillouin zone. As a final interesting point, it should noted in regards of the diagonalization of the 2×2 matrix eigenvalue problem in (10.28) that the unitary matrix which generates the diagonalization now is not a constant matrix. This is unlike the matrix in (10.23) which generated the original block diagonalization made in (10.24).

As an example of the evaluation of (10.29) through (10.33), some results are now presented for an impurity system based on the two-dimensional square lattice model discussed in chapter 3. In figure 10.2 a plot is made of $\delta\varepsilon_{00}$ versus ω for an impurity array composed of the five-site cluster which has just been discussed [4]. The plots in the figure represent the values of the dielectric changes $\delta\varepsilon_{00}$ needed to support an impurity bound state of frequency ω. Here the range of bound state impurity mode frequencies in the drawings are located within the second stop band of the photonic crystal under consideration. The curves shown are mode labeled for the formation of the two distinct s-symmetry impurity bond modes, the two degenerate p-symmetry impurity bound modes, and the single d-symmetry impurity bound mode.

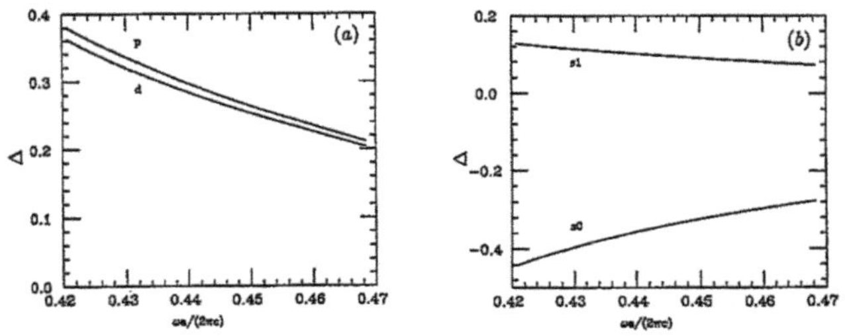

Figure 10.2. Example plots of the value of $\Delta = \delta\varepsilon_{00}S$ needed to support a bound impurity mode versus the stop band frequency ω at which the bound mode occurs. The plots are computed for a particular system which has been treated in the literature [4]. Results are shown for bound state modes of the cluster of five sites having symmetries: (a) p- and d-symmetry and (b) the two different s-symmetries. The plots are given only as an illustration of the theory, and for the details of the system parameters used in the evaluation of the theory the reader is referred to the literature [4]. Reproduced from [4] © IOP Publishing.

The results in figure 10.2 illustrate some of the general features of clusters found within our simple model for impurities formed of linear dielectric impurity media. These results can be extended to handle the case of clusters composed of Kerr nonlinear media. In order to handle weak Kerr nonlinear media, on the left-hand side of (10.29) through (10.31) $\delta\varepsilon_{00}$ is replaced by the Kerr form $\delta\varepsilon_{00}[1 + \lambda\,|E|^2]$ where the field is approximated as that at, for example, the $\lambda = 0$ solution of the cluster. The resulting solution for $\delta\varepsilon_{00}$ needed to support a bound state mode of stop band frequency ω under these considerations now depends on the intensity of the modal field. The discussions are seen to be a direct generalization of those made for the single site Kerr impurity [4].

As the strength of the Kerr interaction is increased, however, a consideration of the field variation over the cluster adds to the complication of the problem. This limit of the problem may require recursive methods based on generating iterative solutions of the nonlinear equations to reach a well-defined solution accounting for modal field variations within the cluster. For further details regarding the treatment of this problem and other aspect of the general study of clusters the reader is referred to [3] and [4].

Finally, it is interesting to note that if the symmetry of the impurity cluster is reduced the degeneracies of the eigenvalues may be changed. Specifically, consider the impurity cluster in figure 10.1(b) and replace the right-hand impurity of the cluster by one which is different. Here the dielectric change in one of the outer cluster sites is taken to be different from the other three outer sites. Now the symmetry group of the cluster is composed of the identity operation in which no change is made and the symmetry involving a reflection through the horizontal axis passing through the center site of the cluster. The total symmetry group of the new configuration is composed of two operations rather than the eight operations of the original symmetry group. In addition, for the new dielectric arrangement there are now only two classes: (1) the class composed of the identity operation, and (2) the class composed of the reflection through the horizontal axis. Since each of these classes contains only a single element of the group, all of the eigenvalues of the cluster of the new group containing only two elements must be singly degenerate, i. e., $2 = 1^2 + 1^2$.

This ability to change the symmetry of an impurity cluster has an interesting technological application. Specifically, some examples of such changes in impurity dielectric have earlier been noted in the dielectric changes induced by Kerr nonlinearity or through mechanisms based on the applications of external electric or magnetic fields. In this respect, if one has a way of changing the dielectric of sites forming the cluster through means of an external chemical, thermal, field interaction or even through some other mechanism internal to the system; this can be used to break the original symmetry of the cluster [9]. As has been seen such changes in the cluster symmetry may ultimately show up as modifications in the degeneracy of the bound state energy levels exhibited by the cluster. These types of changes have been useful in switching or sensor applications which modify the nature of the frequency responses generated by the cluster [9]. Particular examples of chemical and biological sensors based on these ideas are found in the literature [9, 13, 14].

We now turn to the treatment of site coupled photonic crystal waveguides. This forms an interesting case as it allows the waveguides to interact so that information can be transferred from one optical system to another. It also illustrates some interesting effects of nonlinearity as applied in optical circuits.

10.1.2 Waveguides with off-channel interactions

In the next discussions we return to a consideration of single site problems, but now for the case in which the single site acts as a resonant couple, linking the modes of different photonic crystal waveguides. Here the waveguides are formed in the photonic crystal as separate lines of site impurities which are adjusted to bind guided modes at frequencies in the stop bands of the photonic crystal. The coupling sites are off-channel impurity sites which cause a resonant interaction with the modes of the two different waveguide channels.

In this regard, the site provides a means for allowing the modes in two distinct waveguides to interact with one another. It allows for a frequency dependent transfer of modes between the two waveguides, setting up a correlation between the modes in each of the two channels. These types of coordinated units represent essential building blocks for the development of systems of optical circuits [9]. When such units act together their ability to pass energy and information for processing in device technologies is greatly enhanced and is seen to mimic the capabilities for information handling found in existent electronic technologies.

In a first study a consideration is given of the problem of a waveguide formed of linear media which interacts with an off-channel site. The transmission down the waveguide is modified by the scattering with the off-channel site so that the transmission down the channel becomes frequency dependent with a maximum backscattering in the channel occurring at a single resonant frequency. The resonant interaction of the off-channel site allows for specific frequencies passed by the waveguide to be uploaded from or downloaded into the waveguide channel and, by placing a series of different off-channel sites along the length of the waveguide channel, can be used to develop a frequency selector. Through the application of Kerr nonlinear impurities, the field amplitudes of the channel modes can be used to tune the resonant frequencies at which the system operates.

In a second study, two waveguides are coupled through the intermediary of an off-channel site. There is again a resonant frequency at which energy is transferred between the two waveguides. Consequently, while some frequency modes in the two channels are weakly coupled, frequency modes which interact resonantly are strongly coupled between the two guides. The particulars of the resonant interaction are found to be dependent on the dielectric constants and the geometries of the waveguides and the resonant sites. In this respect, a bit of frequency selection is at play in the design of the system. These features of the coupled waveguide system are illustrated by some examples.

10.1.2.1 Waveguides coupled to a single off-channel site

We begin by considering the problem of an infinitely long straight waveguide which is coupled to a single off-channel site [3, 6]. The waveguide is formed in the photonic crystal as a line of site impurities which are adjusted to bind guided modes to them existing in the stop bands of the photonic crystal. Both features are embedded in a two-dimensional photonic crystal defined on a square lattice with nearest neighbor separations along the x- and y-axes. For a convenience, the lengths in the problem are measured in units of the nearest neighbor separations. While the photonic crystal and waveguide channel sites are formed of linear dielectric media, the off-channel feature is composed of a Kerr nonlinear medium [6].

The specific problem of interest is a scattering problem in which a guided mode traveling along the infinite waveguide channel is incident on the off-channel site [6]. Specifically, the focus is on the transmission and reflection coefficients of the elastically scattered guided wave as it interacts with the off-channel feature. The scattering from the feature is studied as a function of the frequency of the incident wave and is of the form of a frequency dependent resonant scattering. Due to the presence of the Kerr nonlinearity the transmission and reflection from the site are also dependent on the electric field intensity of the incident wave.

The propagation in the system described is characterized by a set of nonlinear difference equations. In the description of the infinite waveguide channel, the neighboring waveguide sites are separated by the lattice constant of the photonic crystal. In this regard, the coordinates of the channel sites are along the x-axis at sites $(n, 0)$ where n ranges over the integers. The off-channel site is located at $(0, p)$ where p is a positive integer indicating the number of lattice constants the off-channel site is separated from the waveguide channel. A wave is incident on the left of the off-channel site coming from infinity and generates a reflected and transmitted wave upon passing the off-channel site. (See figure 10.1(c) for a schematic.)

Within the waveguide channel in the regions $|n| > 0$ the difference equations for the propagation within the waveguide are given by [6]

$$E_{n,0} = \gamma\left[aE_{n,0} + b\left(E_{n+1,\,0} + E_{n-1,\,0}\right)\right]. \tag{10.34a}$$

Here a and b are the onsite and nearest neighbor couplings which are ultimately related to the Green functions of the pure photonic crystal, and γ is proportional to the dielectric changes in the photonic crystal responsible for the formation of the waveguide channel. At the $(0, 0)$ site of the waveguide channel there is a coupling to both the $(-1, 0)$, $(1, 0)$ channel sites as well as an additional weak coupling to the off-channel site $(0, p)$. These interactions are given by the form [6]

$$E_{0,\,0} = \gamma\left[aE_{0,\,0} + b\left(E_{1,\,0} + E_{-1,\,0}\right) + cf_\epsilon\left(E_{0,p}\right)\right] \tag{10.34b}$$

where the weak interaction between the channel and off-channel sites c is proportional to the pure photonic crystal Green function between the $(0, 0)$ waveguide site and the impurity site located at $(0, p)$. In (10.34b) the Kerr nonlinear interaction at the off-channel site is characterized by the nonlinear function [6]

$$f_\epsilon(E) = \gamma'[1 + \lambda\,|E|^2]E + i\epsilon E \tag{10.34c}$$

where γ' and λ account for the introduction of Kerr material at the off-channel site, and a small imaginary term $i\epsilon E$ is added to the dielectric so that the system does not become infinite at the resonance. If $i\epsilon E$ is removed from (10.34c) an unphysical singularity occurs in the system. Similarly, the off-channel site $(0, p)$ interacts with the waveguide site at (0, 0) through a difference equation. This interaction is characterized by the form [6]

$$E_{0,p} = af_{\epsilon}(E_{0,p}) + \gamma c E_{0,\,0} \tag{10.34d}$$

which involves both an onsite term which is a function of $E_{0,p}$ and a term dependent on $E_{0,\,0}$ characterizing the interaction with the channel.

To solve the system of difference equations in (10.34) an ansatz is made. Specifically, it is assumed that the solution of the set of difference equations has the general form given by [6]

$$E_{n,0} = ue^{ikn} + re^{-ikn} \text{ for } n < 0 \tag{10.35a}$$

where u represents the amplitude of the incident wave arriving from the left of the impurity and r represents the amplitude of the wave reflected from the impurity;

$$E_{n,0} = te^{ikn} \text{ for } n \geqslant 0 \tag{10.35b}$$

where t is the amplitude of the wave transmitted past the off-channel site; and

$$E_{0,p} = st \tag{10.35c}$$

is the field amplitude at the off-channel site which is complex of the form

$$s = s_0 + is_1 \tag{10.35d}$$

with s_0 and s_1 real.

Upon substitution of (10.35) into (10.34) it follows from some algebra that the transmission coefficient is given by [6]

$$T = \left|\frac{t}{u}\right|^2 = \frac{4\sin^2 k}{\left[2\sin k + \frac{cs_1}{b}\left(1 + 2\frac{b}{a}\cos k\right)\right]^2 + \left[\frac{c}{b}\left(1+2\frac{b}{a}\cos k\right)(s_0 - \gamma c)\right]^2} \tag{10.36a}$$

where s is a solution of the cubic equation

$$\lambda t^2 \,|s|^2\, s + \left(1 - \frac{1 - ia\epsilon}{\gamma' a}\right)s + \frac{\gamma c}{\gamma' a} = 0. \tag{10.36b}$$

At frequencies away from the resonant singularity, in the region in which the dielectric has a negligible effect on the system, the associated reflection amplitude is then given by

$$R \approx 1 - T \tag{10.36c}$$

which is a standard relationship from the definition of R and T.

The expressions in (10.36) have been evaluated for the example of the square lattice photonic crystal treated in chapter 3. In figure 10.3 results are presented in a particular system of parameterization for the resonant scattering from a linear off-channel site

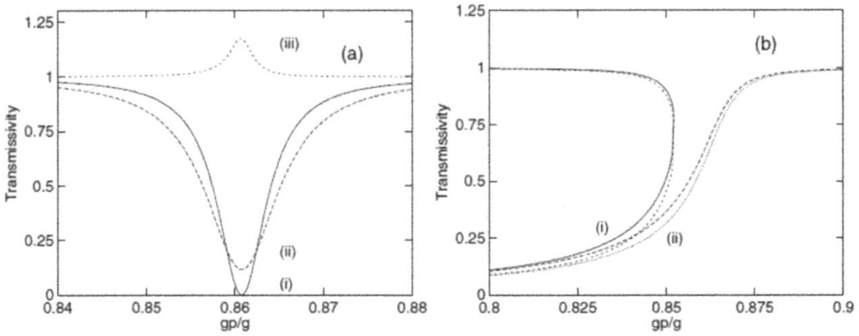

Figure 10.3. Plot of transmissivity versus $gp/g = \gamma'/\gamma$ for transmission through an infinite straight waveguide channel interacting with a single off-channel impurity. The study is made at a fixed incident wave frequency and the resonant frequency of the interaction with the off-channel site is changed as the ratio gp/g is varied. The results are for an off-channel site formed only of: (a) a linear medium and (b) a Kerr nonlinear medium. The curves of interest to us are those labeled (i) and (ii). For the details of the parameterization of the results, the reader is referred to [6]. Reproduced from [6] © IOP Publishing.

and from a Kerr nonlinear off-channel site. The plots present the transmission coefficient or transmissivity as a function of the ratio $gp/g = \gamma'/\gamma$.

Specifically, the transmission in figure 10.3 is for an infinite straight waveguide channel interacting with a single off-channel impurity where the incident wave is set at a fixed frequency. The resonant frequency of interaction with the off-channel site is changed by changing the ratio $gp/g = \gamma'/\gamma$. Consequently, by changing the ratio $gp/g = \gamma'/\gamma$ the strength of the coupling between the waveguide and the off-channel site is varied. The results in figure 10.3 are for an off-channel site formed only of: (a) a linear medium or (b) a Kerr nonlinear medium. For the plots the curves of interest to us are those labeled (i) and (ii). The curves represented in this manner are computed for different dielectric losses. For the details of the parameterization of the results in figure 10.3, the reader is referred to [6.]

An important point to note in the plots is the difference in the nature of the transmission results between the linear and the nonlinear impurity results. Specifically, the nonlinear results exhibit a multiplicity of possible transmissivities for a given ratio $gp/g = \gamma'/\gamma$. This arises from the optical bistability of the Kerr medium and is absent in linear media.

10.1.2.2 Two waveguides coupled through a single site

In a second study, the problem is considered of an infinitely long straight waveguide which is coupled through a single off-channel impurity site of Kerr medium to a semi-infinite waveguide [6]. The two waveguides are oriented so that the channel of the semi-infinite waveguide is directed perpendicular to that of the infinitely long waveguide. In this respect, the new geometry of the system is represented in figure 10.1(d) as a modified version of the problem just treated. As in the earlier consideration all of the system is formed of linear dielectric media except for the single Kerr coupling site. The study focuses on a wave in the infinite waveguide which is incident on the impurity site from its left and in part is transmitted into the infinite and the semi-infinite channels.

As in the previous problem the system treated is based on the square lattice photonic crystal. Again, the infinite waveguide of nearest neighbor sites is situated along the x-axis, with the waveguide site at (0, 0) being coupled to an off-channel site located at $(0, p)$. The site at $(0, p)$ is in turn coupled to the semi-infinite waveguide composed of sites along the y-axis located at the lattice positions $(0, m)$ where $m \geqslant p + 1$ are consecutive integers. In this geometry the semi-infinite waveguide has a channel of the same composition as that of the infinite waveguide along the x-axis, and the coupling site $(0, p)$ is, as in the previous consideration, composed of Kerr nonlinear medium. As a result of the coupling of the Kerr site at $(0, p)$ a resonant transmission is found between the of modes of the two waveguides. For the discussions presented in the follow we take $\epsilon = 0$ so that it is assumed that there are no dielectric losses in the system. For the present problem the introduction of a loss in the system is unnecessary as no singularities arise in the absence of such losses [6].

The problem for the system in figure 10.1(d) is described by the same difference equations as in (10.34a) through (10.34c) for the infinite waveguide coupled to the off-channel Kerr medium site in figure 10.1(c), but must also include some new equations accounting for the added features in figure 10.1(d). Specifically, in addition to (10.34a) through (10.34c), a set of equations accounting for the coupling of the off-channel site to the semi-infinite waveguide need to be introduced into the problem. For these considerations the following difference equations must be included [6]:

$$E_{0,p} = af_0\left(E_{0,p}\right) + \gamma c E_{0,\,0} + \gamma b E_{0,p+1} \tag{10.37a}$$

describing the coupling of the off-channel Kerr site $(0, p)$ to the infinite waveguide at (0, 0) and to the semi-infinite waveguide at $(0, p + 1)$,

$$E_{0,p+1} = \gamma\left[aE_{0,p+1} + bE_{0,p+2}\right] + bf_0\left(E_{0,p}\right) \tag{10.37b}$$

describing the coupling of the semi-infinite waveguide at $(0, p + 1)$ to the Kerr site at $(0, p)$, and

$$E_{0,m} = \gamma\left[aE_{0,m} + b\left(E_{0,m-1} + E_{0,m+1}\right)\right] \tag{10.37c}$$

which describes the semi-infinite waveguide in the region $m \geqslant p + 2$. Notice that in (10.37a) and (10.37b) $\epsilon = 0$ as seen in the form $f_0\left(E_{0,p}\right)$, and this is also the case for (10.34b) and (10.34c) applied to the present problem.

In addition to the ansatz in (10.35) used to treat the system in figure 10.1(c), the present problem now requires an added ansatz component in the form of an outgoing scattered wave traveling away from the coupling in the semi-infinite waveguide channel. This is given by the outgoing wave [6]

$$E_{0,m} = t'e^{ikm} \quad \text{for} \quad m \geqslant p + 1. \tag{10.38}$$

With the introduction of these forms into the set of difference equations and following a bit of algebra, solutions are then obtained for the transmissions and reflections in the system.

In this way it is found for the case of a wave incident from the far-left side of figure 10.1(d) and traveling towards the waveguide coupling that the transmission coefficient of the wave incident on the coupling and passing on through the infinite waveguide is given by [6]

$$T = \frac{4\sin^2 k}{\left|2\sin k - i\frac{\gamma' c}{\gamma b}\left(1 + \lambda t^2 |s|^2\right)s\right|^2}, \tag{10.39a}$$

where s is a solution of the cubic equation

$$\gamma'[a + be^{ik}]\left(\lambda t^2 |s|^2 + 1\right)s - s + \gamma c = 0. \tag{10.39b}$$

On the other hand, the transmission coefficient for a wave incident from the left on the coupling and passing into the semi-infinite waveguide is given by

$$T_{\text{semi}} = \left|\frac{\gamma'}{\gamma}\left(1 + \lambda t^2 |s|^2\right)s\right|^2 T. \tag{10.39c}$$

In the present system there is no loss mechanism so that $T + T_{\text{semi}} + R = 1$ where R is the reflection from the coupling features.

As an illustration of the problem considered in (10.37) and (10.39) involving the coupling of the two waveguides, results are show in figure 10.4 for the transmission

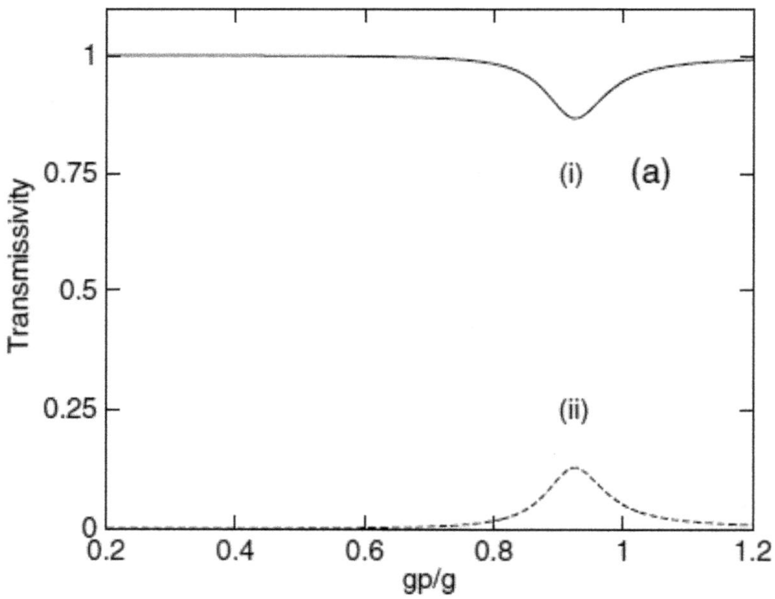

Figure 10.4. Plot of the Transmissivity versus $gp/g = \gamma'/\gamma$ for the two coupled waveguides in the geometry of figure 10.1(d). The solid line is for transmission of the infinite waveguide into itself whereas the dashed line is for the transmission of the infinite waveguide into the semi-finite waveguide. The incident wave is at a fixed frequency and the resonant frequency of the system is changed by changing $gp/g = \gamma'/\gamma$. For the results shown here the system is composed only of linear media. The details of the parameterization can be obtained by consulting [6]. Reproduced from [6] © IOP Publishing.

of a wave in the infinite waveguide incident on the coupling. The example system involves the same square lattice photonic crystal as treated in the earlier considerations of the resonant scattering from the single off-channel Kerr nonlinear medium site. In the present plot, however, only results for the case in which linear media occurs in the system are presented.

Both transmission coefficients (i.e., the transmissivities) for a fixed frequency incident wave propagating past the coupling and scattered into the infinite channel and into the semi-infinite waveguide are plotted versus $gp/g = \gamma'/\gamma$. As in the system discussed earlier in figure 10.1(c) the resonant coupling between the two waveguides is changed by varying $gp/g = \gamma'/\gamma$. The transitions between the two waveguides show various resonant features depending on the parameterization of the system, but there is no bistability in the absence of Kerr media. For further detains, the reader is referred to the literature [6].

The considerations of this section have involved the study of single sites, clusters of sites, and couplings between waveguides. These types of problems are of ordered geometries which can be of application in the design of optical circuits, e.g., in the formation of optical resonators and wave-guided components. We next consider the treatment of truly random disordered materials which involve elements of statistically characterized disorder throughout the entire system. These types of considerations are of importance in the study of uncertainties occurring in the manufacturing processes, but they may also be of importance in their own right as in the general theory of Anderson wave localization. In fact, optical systems have become particularly important in the study of Anderson localization itself.

Specifically, we shall now look at the basic ideas of Anderson localization which involve fundamental issues regarding the nature of the propagation of wave excitations in randomly disordered materials. In this study, it is found that statistically random disorder can change the basic nature of the wave functions of the modes in the system from being extended throughout the entire medium to being localized within finite regions of the medium. This transition is an example of a phase transition in which a medium changes from one that transports energy to one that does not. The driving mechanism in the transition is the amount of disorder in the system.

10.2 Anderson localization

In uniform or periodic media, the wave functions are extended throughout the uniformity or periodicity of the material [1, 7–12]. This follows from our earlier discussions of the plane wave or Bloch wave nature of the modes and arises from the translational symmetry properties of the system. With the introduction of random disorder into a material its translational symmetry is destroyed. If the disorder is weak the wave functions remain extended over the material, but if the disorder is strong enough a new type of wave function can be introduced into the modal solutions. These new solutions are the so-called localized modes which, unlike extended modes, are spatially confined to a finite subregion of the disordered medium and are typified by wave function envelopes of an exponential form [1, 7–12], i.e.,

$$\psi(\vec{r}) \propto e^{-\frac{|\vec{r}|}{l}}. \tag{10.40}$$

Here $\vec{r}$ is a spatial vector from the center of the localized wave function and l, known as the localization length, characterizes the radius of the region in which the wavefunction is localized. It is seen that in the limit that $l \to \infty$ the localized modes in (10.40) transform into extended modes as the system regains its translational symmetry.

The nature of these two different types of wave function solutions in a material greatly influences the transport and scattering properties it exhibits [1, 7–12]. An important example of this influence on transport is the metal–insulator transition in disordered electronic media. In the metal–insulator transition it is found that with increasing disorder a point is reached in metallic materials in which, due to the introduction of localized modes in the system, a metallic material changes abruptly from a conductor to an insulator. This transition in the transport properties, however, is not limited to electrical conductivity.

As the origins of extended and localized modes are fundamentally related to the disorder in the system, more recently studies have shown similar types of reductions in the transport of photons and phonons in disordered optical and acoustic media [1, 7–12]. These again arise through the same type of transport mechanisms as those found in the metal–insulator transition. In all of these systems the effect on transport comes from the localization properties of the wave functions. Specifically, only the extended modes of a system carry energy through the material. The localized modes do not span the entire extent of the material and at zero temperatures they are unable to pass a current out of their region of localization. At nonzero temperatures, however, the localized modes only limit the transport of energy through the system as thermal hopping between localized modes occurs.

In addition, as the system enters a localized state the excitations introduced into it encounter an enhanced backscattering as they move through and are elastically scattered by the medium [1, 7–12]. This arises from the phase information carried by the scattered waves and is present in all types of disordered media whether electronic, optical, or acoustic. In fact, the transition to localized modes is driven by the degree of enhanced backscattering present in a material. A particularly important example of enhanced backscattering in optics is observed as an intensity peak in the angular distribution of diffusely scattered light by a disordered medium in the backwards direction. In these systems the media enhances the elastically scattered light traveling opposite to the light in the incident beam, and this also limits the penetration of light into the medium.

The transition of wave functions from extended to localized modes with the introduction of disorder into a system is known as Anderson localization, and it is a property arising from the wave nature of the modes and their elastic scattering arising from the disordered medium [1, 7–12]. It is independent of whether or not the systems studied are classical or quantum mechanical in nature but is primarily a function of the degree of scattering encountered in the system. In this regard, at the localization transition a balance is reached between the desire of the system to propagate a plane wave and the intensity of the disorder limiting this propagation.

Consequently, an early proposed criterion for understand the onset of the transition from extended to localized wave functions is the Ioffe–Regel criterion [1, 7–12]. This provides a rough guide as to when modes should be extended and when they should be localized in space. In understanding modal transport properties, the extended wave nature of a transport mode of the system is characterized by the wave number k, related to the rate it transports energy in the system, and the mean free length l_0 characterizing the distance traveled between successive scatterings. As the elastic scattering intensifies l_0 decreases and the wave number k becomes less well defined as a characterizing property of the wave functions.

A question then is when does the description of the wavefunction by k begin to become meaningless. One would surely expect that this might occur when the mean free path for scattering is of order of the wavelength of the mode. At this point scattering effects are beginning to dominate the system so that scattering and propagation are becoming equally important in characterizing the modes. Combining the wave number of the extend state behavior and the scattering length of the localizing effects, the boundary between extended and localized modes should occur when

$$\mathrm{kl}_0 \sim 1. \tag{10.41}$$

This is the Ioffe–Regel condition at the boundary of the localized–extended state region. In the case that $\mathrm{kl}_0 < 1$ scattering dominates and the mode is localized while in the case that $\mathrm{kl}_0 > 1$ the scattering has less effect on the mode and the mode becomes an extended state.

In the following a discussion of the basic properties of localization effects is presented. Our interest ultimately is in optical effects associated with photonic crystals and metamaterials. These systems have numerous complications associated with the polarization properties of optical waves and their dispersive properties. As a consequence of this and of the history of the development of the subject [1], however, first some of the localization properties of electron systems will be discussed. The treatment and understanding of localization in electron system is much simpler than in optical systems, and electron properties have been treated much more extensively than those of optical materials. This will greatly facilitate the understanding of localization in the optical systems, the discussion of which will then follow the treatment of electron localization.

10.2.1 Simple treatment of electron localization

In the discussions of electronic transport presented in elementary physics courses a distinction is made between resistance and resistivity [1, 7–12]. The resistance is a property of a device and depends both on the nature of the material composing the device and on its geometry. In this regard, an example of resistance is a uniform length of copper wire having a specific cross sectional area. On the other hand, the resistivity is a property of a material arising from its specific atomic makeup. An example is the bulk resistivity of copper which characterizes the ability of bulk copper to transport electrons.

The bulk resistivity depends on temperature, impurity concentration, and the application of external fields. For most engineering applications the resistivity is treated as the bulk resistivity which is independent of the sample size. As shall be seen later this last point is an assumption which needs further consideration in some nanoscience applications. This occurs especially for the cases of small samples of materials and for systems at very low temperatures.

The resistance and resistivity are related to one another through geometric considerations. In the model of a uniform length of wire in three-dimensions the resistance, R, of the wire is given by [1, 7–12]

$$\mathrm{R} = \frac{L}{A}\rho \tag{10.42}$$

where L is the length of the wire, A is its cross sectional area, and ρ is the bulk three-dimensional resistivity of the material composing the wire. Consequently, for a three-dimensional cube of material, it follows that $A = L^2$ so that (10.42) becomes

$$\mathrm{R} = L^{-1}\rho. \tag{10.43}$$

In two-dimensions the resistor is a film characterized geometrical by a length L across which a potential is applied and a width W perpendicular to L. If $L = W$, then the resistance is written as

$$\mathrm{R} = \frac{L}{W}\rho_2 = \rho_2 \tag{10.44}$$

where ρ_2 is the bulk resistivity of the material formed as a film. An example of such a system is a plane of atoms or a two-dimensional election gas in a MOSFET transistor. Similarly, in one-dimension only the length L of the material enters into determining the resistance, and

$$\mathrm{R} = L\rho_1 \tag{10.45}$$

where ρ_1 is the bulk resistivity of an infinite line of material. An example of such a one-dimensional system would be a chain of atoms, a nanowire or any type of whisker [1].

In summary, we find that the resistance in d-dimensions, R_d, is given by [1, 7–12]

$$\mathrm{R}_d = L^{-(d-2)}\rho_d \tag{10.46}$$

where ρ_d is the bulk resistivity of the material treated and $d = 1, 2, 3$ labels the dimensionality of the resistivity. Notice that for the same material the bulk resistivity in one-, two-, and three-dimensions can be different.

Early theories of the resistivity studied the transport properties of electrons in terms of classical mechanics. In these approaches the electrons are regarded as charged point masses which do not interact with one another, but which do interact with a viscous force representing the scattering of the electrons in the transporting medium. (The development is similar to that for the terminal velocity of a falling mass in the atmosphere.) From these simple considerations the resistivity of a d-dimensional bulk material in an applied electric field is shown to be of the form [1, 7–12]

$$\rho_d = \frac{1}{en_d\mu} \tag{10.47}$$

where e is the unit of electrical charge, μ is the electron mobility characterizing the inertial properties of an electron, and n_d is the d-dimensional carrier density of the electrons in the material. Since the electrons are regarded as point masses in a bulk medium the resistivity ρ_d is independent of the size of the sample of material being studied. However, with the introduction of quantum mechanics into the considerations and the treatment of the electrons as waves, this independence may no longer be the case.

Upon the introduction of the wave nature of the electrons into the problem, the resistance of a sample of the conducting material is no longer represented by (10.46). The electrons now provide a new length scale to the system which enters the problem as a modifying factor. Specifically, the electrons carry a wavelength λ obtained from their Schrödinger equation solutions, and this affects both quantitatively and qualitatively the nature of the electron transport. The new information provided by the wavelength changes the physics of small material samples and general sized material samples at very low temperatures.

Treating the bulk resistivity in the model in which the electrons are classical mechanical point particles only requires the introduction of the mean free path l_0 as a length characterizing the scattering. In this regard, the mean free path sets the magnitude of the mobility in (10.47) and the resulting value of μ is the same throughout the medium. With the introduction of quantum mechanics to the problem the electron becomes a wave of wavelength λ, and its scattering is now characterized by both the l_0 and λ lengths. However, from the earlier discussions of the Ioffe-Regal criterion in (10.41) it is seen that these two lengths provide the condition for determining the breaking point between the region of extended and localized wave functions in the system.

Specifically, in the case that $kl_0 > 1$ the scattering is not a sufficiently large effect on the propagating modes to change their transport properties, and the modes are characterized as extended states. Consequently, in this limit the disorder in the medium only affects the value of the mobility in (10.46) and (10.47). On the other hand, in the case that $kl_0 < 1$ scattering dominates the modes to the extent that they transform to become localized modes. Now, in this strong scattering limit, the actual nature of the wave functions is changed, and from (10.40) it is seen that a new length scale, the localization length l, enters the problem. This localization length introduces an important new feature in a variety of engineering considerations and applications.

In the limit of weak scattering, it is found within a leading order perturbation treatment of the quantum mechanical electron impurity scattering that localization does not enter into determining the form of the resistivity of samples of materials. In this regime of lowest order scattering an Ohms law result (i.e., the independence of the resistivity on the length of the sample) is obtained in each dimension. Consequently, from (10.46) the dependence of R_d on the length L of the sample of material is given from (10.46) by the proportionality [1, 7–12]

$$\mathrm{R}_d \sim L^{-(d-2)}, \tag{10.48}$$

indicating a power law dependence on L.

On the other hand, in the other extreme limit of very strong scattering in which the conduction electrons in the material are becoming localize, the electrons are characterized by localized wavefunction envelopes of the form represented in (10.40). All of the electron wave functions in this limit are approaching localized forms having a spatial dependence typified by exponential tails in the separation from the center of gravity of the electron wave function. As a result, the movement of electrons through the material would have to be supported by electron hopping or tunneling between neighboring localized modes where the tails approach or overlap. In this picture, the rate of tunneling would have to be dependent on the exponentially small hopping and would yield a dimensionally dependent resistance characterized by a sample length dependence of the form [1, 7–12]

$$\mathrm{R}_d \sim e^{A_d \frac{L}{L_0}}. \tag{10.49}$$

where $A_d > 0$ is a dimensionally dependent constant and L_0 is a constant characteristic length of the system. (More will be said of the nature of L_0 later in the context of a minimum electrical conductance. For now, it may be thought of as a constant value used to make $\frac{L}{L_0}$ in equation (10.49) dimensionless.) For this limit the resistance now depends exponentially on L. In both limits, however, an important point is that at fixed temperature, in the absence of applied magnetic fields, and for fixed chemical composition, R_d only depends on L by the proportionalities in (10.48) and (10.49).

Assuming zero temperature, the absence of applied magnetic fields, and a fixed chemical composition (10.48) and (10.49) describe the resistance of a sample of material as a function of L. The question of interest now is: how does the resistance of the sample behave between the two opposite limits of extremely weak and extremely strong electron scattering? A way to approach this problem (which was developed early in the study of electron localization) is in the context of a scaling theory treatment.

10.2.1.1 Scaling theory treatment of the resistance

Scaling theory has its origins in ideas developed for the study of phase transitions, and the metal–insulator problem is a type of phase transition [1, 7–12]. In addition, the ideas of the scaling theory of resistance which shall now be presented have ultimately been supported by computer simulation studies as well as by experiments [1]. As an important point for the initial studies, one of the conditions on the presentation is the absence of applied magnetic fields. The reasons for this restriction, as we shall briefly elaborate upon later, is that magnetic fields break time reversal symmetry, and this fundamentally changes the problem. Consequently, early discussions of the scaling theory of resistance did not treat applied magnetic fields [1], and the scaling theory now discussed does not include effects of applied magnetic fields.

To initiate the scaling approach, begin by taking the logarithm of (10.48) and (10.49). In the case of (10.48) this gives [1]

$$\ln \mathrm{R}_d = (2 - d)\ln L, \tag{10.50}$$

and in the case of (10.49) it follows that

$$\ln \mathrm{R}_d = \ln\left[e^{A_0 \frac{L}{L_0}}\right] = \frac{A_d}{L_0} e^{\ln L}. \tag{10.51}$$

These two limiting forms represent $\ln \mathrm{R}_d$ as a function of $\ln L$ which is just an alternative way of saying that R_d is a function of L. Nevertheless, taking the derivative of $\ln \mathrm{R}_d$ with respect to $\ln L$ generates a very important expression for use in the following discussions.

In this way it then follows from (10.50) that

$$\frac{d \ln \mathrm{R}_d}{d \ln L} = 2 - d, \tag{10.52}$$

and from (10.51) that

$$\frac{d \ln \mathrm{R}_d}{d \ln L} = \frac{A_d}{L_0} e^{\ln L} = \frac{A_d}{L_0} L. \tag{10.53}$$

Or alternatively viewed in terms of R_d and L these are just

$$\frac{dR_d}{d\mathrm{L}} = (2 - d)\frac{R_d}{L} \tag{10.54}$$

from (10.52) and

$$\frac{dR_d}{d\mathrm{L}} = \frac{A_d}{L_0} R_d \tag{10.55}$$

from (10.53).

These first order differential equations are for the known limits on the system [1]. Of interest now is how these limits are connected up for intermediate degrees of electronic scattering by the disordered medium. This is a very difficult problem, but by giving a little thought to the physics of the problem an insight is developed into the qualitative nature of the connection between the two limiting behaviors. In this regard, the simplest treatment of the system is to consider that in the intermediate regions the derivatives in (10.52) through (10.53) only depend on L or alternatively only on the functionally related R_d. Really, we have no reason to argue a more complicated behavior for the system which might include other variables of the material, and this viewpoint will be adopted in the following.

As a reasonable assumption it is now postulated that the correct generalization of (10.52) through (10.55) in the known weak and strong scattering limits and throughout the region of intermediate scattering is given by the general forms [1]

$$\frac{d \ln \mathrm{R}_d}{d \ln L} = \beta_d(L) = \beta_d(\mathrm{R}_d) \tag{10.56a}$$

or alternatively by

$$\frac{dR_d}{d\mathrm{L}} = \frac{R_d}{L}\beta_d(L) = \frac{R_d}{L}\beta_d(\mathrm{R}_d). \tag{10.56b}$$

Here $\beta_d(L)$ is a function which depends on the dimension d of the system and on the length of the system L, and alternatively $\beta_d(\mathrm{R}_d)$ is obtained from $\beta_d(L)$ through the functional dependence of R_d on L. In addition, some other assumptions can now be placed on the behavior of $\beta_d(L)$ and $\beta_d(\mathrm{R}_d)$ arising from general considerations of the nature of the resistance of materials. These are now discussed.

For example, in materials in which the modal wave functions are transforming or have transformed into localized modes, it is expected that $\beta_d(L)$, $\beta_d(\mathrm{R}_d) > 0$. In this region and regarding the $\beta_d(L)$ form of the derivative equality, the resistance R_d is seen from (10.56) to increase with increasing length L. Specifically, when the length of the sample is increasing but still smaller than the localization length characterizing the spatially confined wave functions (i.e., when $L < l_d$ where l_d is the localization length of the d-dimensional system), the system is less dependent on hopping transport and exhibits only a small increasing resistance. However, for increasing length in the region $L > l_d$ the resistance arises from hopping between wave functions of the sample so that it increases much more quickly.

As an example to illustrate this point, in the case that $\beta_d > 0$ is a length independent constant it follows from (10.56) that [1, 7–12]

$$R_d \sim L^{\beta_d} \tag{10.57}$$

causing the resistance to increase with increasing sample size as a power law. Notice, however, that the form in (10.57) eventually falls short of the much more rapid exponential divergence of the resistance with length postulated in (10.49). Consequently, to arrive at the limiting behavior in (10.49) it is reasonable to assume that $\beta_d(L) > 0$ is a monotonically increasing function of L. This accounts for the observed exponential divergence of the resistance with increasing length. In this regard, R_d is functionally related to L so that it follows in this region that $\beta_d(R_d) > 0$ and it is expected that $\beta_d(R_d)$ increases monotonically with increasing resistance.

On the other hand, in materials in which the modal wave functions under increasing length are transforming between extended modes, it is possible that $\beta_d(L)$, $\beta_d(R_d) < 0$.In this region, the resistance R_d as seen from (10.56) tends to decrease with increasing length L. Specifically, as the length increases the geometric factors dominate over the much weaker length dependent effects on the resistivity arising from the renormalization of the extended mode wave functions.

As an illustrative example of this case, in a system for which $\beta_d < 0$ is a length independent constant it follows from (10.56) that [1, 7–12]

$$R_d \sim L^{-|\beta_d|}. \tag{10.58}$$

From (10.58) the resistance is found to decrease with increasing sample size. This behavior is consistent with the limiting form in (10.48) for the specifics of a system in which

$$2 - d < 0. \tag{10.59}$$

As with the earlier discussions of $\beta_d(L) > 0$, it is expected that $\beta_d(L) < 0$ is an increasing function of length. Specifically, as the resistance R_d is increased it is expected that $\beta_d(R_d) < 0$ increases toward the positive region of increased disordered. Again, it is important to remember for making all these transformations that R_d is functionally related to L.

In conclusion, for the purposes of discussion it is assumed that $\beta_d(R_d)$ is generally an increasing continuous function of R_d. This has been justified by computer simulation and experimental studies [1, 2]. The functional dependence of $\beta_d(L)$ on L, however, is more complex and depends on whether the modes considered are extended modes for which $\beta_d(L) < 0$ or localized modes with $\beta_d(L) > 0$. This behavior is now explained.

Taking a closer look at the condition (10.59) it is seen that the condition $2 - d < 0$ is really only satisfied by $d = 3$. Consequently, for $d = 1$ and 2, it follows that $2 - d \geqslant 0$ and due to the monotonic increasing nature of $\beta_d(R_d) > 0$ these systems should always be localized. In fact, this result is borne out by studies of the quantum mechanical perturbation corrections to the weak scattering (Ohmic resistance) limit upon which the condition is based. From these studies it is found that, indeed, the corrected resistance for the $d = 1$ and 2 systems always increases with increasing L over that of the uncorrected resistance. Consequently, $d = 1$ and 2 systems are always localized.

The reason, however, that localization is not observed experimentally in many one- and two-dimensional materials is that in these systems the localization length can be much greater than the sample size [1, 7–12]. For such systems the increase in resistance may be dominated by the leading order Ohmic resistivity, with the contribution from the localization corrections generally too small to matter.

For $d = 3$, however, a material can exhibit either extended mode transport or localized mode behavior. In these materials the determining factor for the transport displayed is the position of the Fermi energy of the system in the energy excitation spectrum of the material. This follows as it is known that at low temperatures the electrons responsible for the transport properties of a medium are located at its Fermi energy. Consequently, it is found that if the Fermi energy is located within the region of localized modes the material exhibits the insulating behavior of localized modes, and if the Fermi energy is located within the region of extended modes the material exhibits the conductive transport of extended modes. In this regard, the energy separating the energy regions of localized and extended modes is known as the energy of the mobility edge.

10.2.1.2 Minimum conductivity in two-dimensions

A final interesting point of the dependence of the resistivity on disorder is in regard to the electrical conductivity of the material, where the conductivity σ is related to the resistivity by $\sigma = \frac{1}{\rho}$. Specifically, a question about the properties of the system considered as a function of the system disorder is: What is the conductivity at the point of disorder in the material at which the metal–insulator transition occurs? This can be found by considering the Ioffe–Regel criterion in (10.41) which sets general conditions of the system at the transition point.

From elementary physics courses the conductivity of a material carrying an electrical current density is given by the simple formula

$$\sigma = \frac{ne^2\tau}{m} = \frac{ne^2}{m}\frac{l_0}{v_f} = \frac{ne^2 l_0}{\hbar k_f}. \tag{10.60}$$

Here n is the carrier density, m is the effective carrier mass, e is the unit of electrical charge, l_0 is the mean free path of the charge carriers, v_f is the Fermi velocity of the carriers, $\tau = \frac{l_0}{v_f}$ is the mean time between successive scatterings in the medium, and $\hbar k_f = mv_f$ for the Fermi wave number k_f is the effective momentum of the carriers. The expression in (10.60) can now be additionally simplified by noting that regarding the nature of n it is a result in basic quantum mechanics that for a d-dimensional material the carrier density is related to the Fermi wave number by

$$n \propto k_f^d, \tag{10.61}$$

and consequently from (10.60) it follows that at the metal–insulator transition [1]

$$\sigma = \frac{ne^2 l_0}{\hbar k_f} = \frac{e^2}{\hbar}\frac{n}{k_f} l_0 \propto \frac{e^2}{h} k_f^{d-2}. \tag{10.62}$$

Note that in arriving at (10.62) the Ioffe–Regel criterion $k_f l_0 \sim 1$ is also applied in the last proportionality in (10.62) to eliminate l_0, so that the far-right expression then relates the conductivity to the Fermi wave number at the metal–insulator transition.

Note from (10.62) that in the case of two-dimensions the minimum conductivity at the point of the metal–insulator transition is solely dependent on the fundamental constants e and $\hbar$. Specifically, the magnitude of the conductivity at the transition is given by

$$\sigma_{\min} = \frac{e^2}{\hbar}, \tag{10.63}$$

so that (10.63) then represents the smallest conductivity the material can have before it becomes an insulator. It follows from (10.46) that a useful unit of measure of the resistance of a two-dimensional material is then given by $R_2 = \rho_2 = \frac{1}{\sigma_{\min}}$, and at the metal–insulator transition of a two-dimensional material

$$R_2 = \frac{\hbar}{e^2}. \tag{10.64}$$

The result in (10.64) is important as a general characterization of the resistance and has found application in the presentation of computer simulation and experimental data of electrical transport properties [1.2]. Specifically, in the study of the resistance with length in a general d-dimensional system a common usage is to present the system transport properties as plots of $\beta\left(\frac{R_d}{R_0}\right)$, $\ln\left(\frac{R_d}{R_0}\right)$, and $\ln\left(\frac{L}{L_0}\right)$ where often $R_0 = R_2$. This enables the study of the relationship between sets of dimensionless parameters of a system. (Note: in this regard, earlier in (10.49) the length L_0 was introduced as

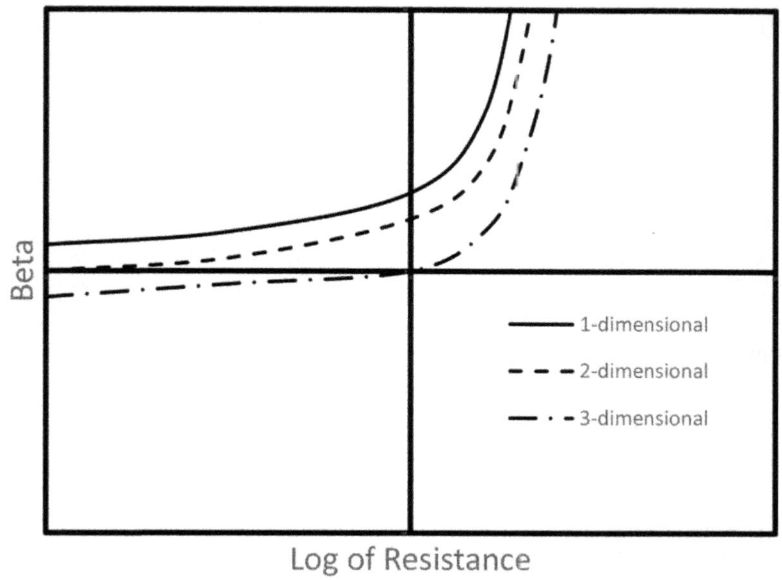

Figure 10.5. A schematic sketch of $\beta\left(\frac{R}{R_0}\right)$ versus $\ln\frac{R}{R_0}$ for 1-, 2-, and 3-dimensional systems. Notice that between the $\ln\frac{R}{R_0} \to \pm\infty$ limits the curves are speculative, being based on the fact that $\beta\left(\frac{R}{R_0}\right)$ is a monotonically increasing function with increasing $\ln\frac{R}{R_0}$. Consequently, the horizontal and vertical lines represent the x- and y-axes intersecting at the origin but are otherwise not scaled. At $\ln\frac{R}{R_0} \to -\infty$ the function $\beta\left(\frac{R}{R_0}\right) = 2 - d$ where d is the dimension of the material, and at $\ln\frac{R}{R_0} \to \infty$ the function $\beta\left(\frac{R}{R_0}\right)$ diverges. Note that, from the definition of R_0, the two-dimensional curve passes through the origin.

characterization of the lengths in the scaling system.) The details of this will be discussed later in the section on simulation studies of optical and electron systems.

For now, in figure 10.5 a schematic figure summarizing the earlier discussions of the properties expected in a plot of $\beta\left(\frac{R_d}{R_0}\right)$ versus $\ln\left(\frac{R_d}{R_0}\right)$ is shown. In the figure the weak and strong scattering limits are shown, and the dotted lines are hypothetical behaviors as discussed earlier. Such a plot as this has been validated by both computer and experimental studies [1, 2].

So far, the nature of the phase coherent scattering which gives rise to the transformation from extended modes to localized modes of the material has not been explained. The particular processes involved in this transformation of the modes arise from a specific subset of all of the scattering processes in the disordered medium. This subset, existing solely because of the wave nature of the excitations, is now discussed.

10.2.2 Phase coherent scattering

The nature of the elastic scattering of a wave by a uniformly disordered medium is intimately connected with whether the wave exists as an extended transport mode of the medium or as a localized non-transporting mode [12, 15]. For weak scattering the modes of a medium are generally extended states, and as the scattering increases in strength a point is eventually reached in which these states would become localized. Ultimately, at a given frequency with a given degree of disorder either all of the modes at that frequency are localized or all of them are extended, but for a given modal frequency these two types of states do not simultaneously exist within the same uniformly disordered medium. In this regard, if there were simultaneous localized and extended states at the same frequency in a medium, they would overlap one another so as to combine to form only an extended mode of the system.

Not all scattering processes contribute to the formation of localized modes [7–12, 15]. Specifically, localized modes are found to have a high degree of contribution to their dynamics from so-called double passage scattering effects. These are effects that were originally introduced in the studies of the optical glory and more recently in the optical treatment of enhanced retroreflection from disordered bulk media and surfaces [15]. They are based on the set of multiple scattering processes along extended paths in the disordered medium which represent pairs of time-reversed trajectories. The double passage effects, depending on their strength, give rise to the weak and strong localization phenomena to be discussed later.

As an example of a set of double passage processes consider the scattering in figure 10.6, representing the passage of two different rays through a randomly disordered medium consisting of five scattering sites [15]. A first ray of light is scattered sequentially by the sites labeled 1, 2, 3, 4, 5. It is first incident on site 1 and eventually has a final scattering on site 5. For the arguments to follow the light is represented as a ray which carries with it phase information. In addition to the scattering within the cluster, the incident light on site 1 and the final outgoing ray from the scattering at site 5 travel anti-parallel to one another. This process represents a retro-reflectance from the medium.

Now consider a second ray passing through the array of sites in figure 10.6 which travels along the time-reversed trajectory to that taken by the first ray. For this

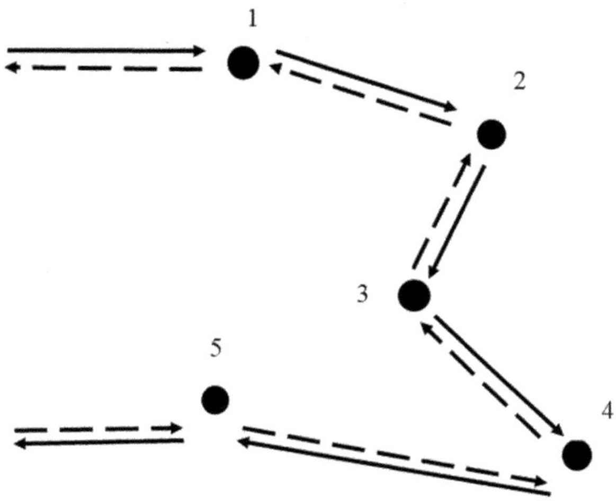

Figure 10.6. Schematic diagram of the double passage effect. The scattering of 1, 2, 3, 4, 5 and the scattering of 5, 4, 3, 2, 1 are time-reversed passages through the medium which, consequently, add phase coherently to give an enhanced contribution to the intensity of retroreflected light from the disordered medium.

second ray the incident ray moves anti-parallel to the final scattered first ray in our earlier consideration and is now incident on site 5. The second ray is then scattered sequentially from 5, 4, 3, 2, 1, emerging from its final scattering at site 1. After its final scattering by 1 it travels anti-parallel to the incident first ray of our earlier considerations. It represents another retroreflection process from the array of sites, this time from the passage of the second ray. The amplitude of the ray of the second process now combines with that of the first process to represent a net retroreflection of both rays from the medium.

If the scattering amplitude for the consecutive scattering of the first wave through the sequence 1, 2, 3, 4, 5 of sites is given by [15]

$$Ae^{i\varphi} \tag{10.65a}$$

where A is the real amplitude of the passage and φ is its phase, then for the reversed scattering sequence 5, 4, 3, 2, 1 passage of the second wave through the system the amplitude is given by

$$Ae^{-i\varphi}. \tag{10.65b}$$

The total scattering amplitude of the medium is the sum of all the possible distinct amplitudes, but to understand the nature of phase coherence consider only the two processes in (10.65). For the case in which only the two amplitudes in (10.65) are involved, it follows that the total scattering amplitude is

$$A_{\text{total}} = Ae^{i\varphi} + Ae^{-i\varphi}. \tag{10.66}$$

The intensity from the total scattering amplitude in (10.66) is then [12, 15]

$$I \propto |Ae^{i\varphi} + Ae^{-i\varphi}|^2 = 4\,|A|^2. \tag{10.67}$$

An important point to note here is that the phase does not enter into the intensity in (10.67). Consequently, if the five sites in figure 10.6 are averaged over a statistical distribution for their positions in space [15]

$$\langle I \rangle \propto 4\,|A|^2 \tag{10.68}$$

where the notation $\langle\rangle$ indicates an average over a random set of spatial position of the sites 1, 2, 3, 4, 5.

Now consider two other arbitrary scatterings of light from the five sites taken in sequences which are random permutations of 1, 2, 3, 4, 5, but having the same initial and final scattering encounters and which do not include the time reversal effects discussed earlier. Specifically, the directions of the initial incident and final scattered rays are the same for each of these two scattering process. (For example, one scattering might first encounter 2 followed consecutively by the sites 3, 4, 1, 5 with 5 being the last encounter, and the other scattering might first encounter 2 followed consecutively by 4, 1, 3, 5 with 5 being the last encounter.) These can be represented by a first process, going from an initial site i to a final site j, of scattering amplitude [15]

$$Be^{i\varphi} \tag{10.69a}$$

and a second process, going from site j to site i, of scattering amplitude

$$Ce^{i\vartheta}. \tag{10.69b}$$

Here B and C are the real amplitudes and φ and ϑ are phases generated upon the passage. Now the amplitude associated with the combination of the two processes in (10.69) is

$$A_{total} = Be^{i\varphi} + Ce^{i\vartheta} \tag{10.70}$$

and the associated scattering intensity is

$$I \propto |Be^{i\varphi} + Ce^{i\vartheta}|^2 = |B|^2 + |C|^2 + 2BC\cos(\varphi + \vartheta). \tag{10.71}$$

If the average of all possible random spatial configurations of the five sites is made, it then follows that $\langle\cos(\varphi + \vartheta)\rangle = 0$ so [15]

$$\langle I\rangle \propto |B|^2 + |C|^2 \quad . \tag{10.72}$$

Unlike in our earlier discussions of time reverse paths, the results for non-time reversed paths exhibit phase effects which upon averaging decrease their contributions to the net intensity.

Similarly, it is found that the result in (10.69) through (10.72) follow for other paths involving the five sites which are not subject to time reversal effects, including those which involved different initial and final scattering sites. For all these scattering processes the key issue is the intensity reduction from the phase averaging.

In the special case that $A = B$ equation (10.72) reduces to [15]

$$\langle I\rangle \propto 2\,|A|^2, \tag{10.73}$$

and a comparison can be made with the earlier time reverse scattering result in (10.68). In this regard, comparing the intensity in (10.68) with that in (10.73) for retroreflected rays, it is found that the intensity of the time reversed phase coherent result in (10.68) is twice that of the retroreflection result for the general incoherent paths in (10.73). The conclusion is that there is an enhancement arising from phase coherent effects in the diffuse scattering of radiation from a random system. This enhancement is observed as a narrow peak in the diffuse scattering intensity as a function of scattering angle which is centered about the direction anti-parallel to the incident beam. This intensity peak is a manifestation of a phenomenon known as weak localization. It is found in the diffuse scattering of light from bulk random medium and from random rough surfaces and is also observed in electron and acoustic scattering from disordered systems. It is a property arising solely from the wave nature of the excitation.

In weak scattering systems weak localization shows up as an enhancement in the retro-reflectance from or within the medium. It is an opposition which acts as a drag on the motion of the wave excitation but does not fundamentally change the nature of the modal wave function. With an increase in the scattering in the medium, however, the backscatter increases to a point at which a transition to the so-called strong localization limit is made. For this limit the wave excitation is confined to be a localized mode. In this regard, the resulting localized mode finds that in every direction it tries to move it is opposed by a strong resistive backscattering. The location of the transition points at which strong localization occurs is highly dependent on the dimensionality of the system, the form of the wave equation for propagation in the random medium, and the values of the parameterization of the wave equation. This difference shall now be addressed for the cases of the modes of the Schrödinger and Helmholtz types of wave equations in disordered media.

10.2.2.1 The difference between Schrödinger and Helmholtz systems: periodicity with disorder

First consider an electron gas in the absence of a periodic potential and the Ioffe–Regel condition for localization of the electron gas. For the case of the electron systems discussed earlier, the wave equation was of the general Schrödinger form [12, 15]

$$[-\nabla^2 + V(\vec{r}) - \epsilon]\psi(\vec{r}) = 0 \tag{10.74}$$

where $\psi(\vec{r})$ is the wavefunction, $V(\vec{r})$ is the random potential describing the electron interaction with the medium, and ϵ is the energy eigenvalue. From the discussions of the Ioffe–Regel criterion in (10.41) it is found that for

$$\mathrm{kl}_0 = 2\pi\frac{l_0}{\lambda} < 1, \tag{10.75}$$

where k is the wave number of the plane wave modes of the $V(\vec{r}) = 0$ gas, the scattering dominates in the random system and the electron states become strongly localized. In the context of the electron gas model of conduction, the material undergoes a metal–insulator transition when the excitations at the Fermi surface transform into localized modes. For this to occur, the wave number k in the Ioffe–Regel criterion must be located at the Fermi surface of the gas.

Now for the case that the potential $V(\vec{r})$ is composed of a component from a periodic ion lattice along with a weaker added component of randomness, the wave functions of the electrons in the periodic lattice dominate the discussions. The modes from the periodic lattice part of the potential are of the Block form

$$\psi(\vec{r}) = e^{i\vec{k}\cdot\vec{r}}u_{\vec{k}}(\vec{r}) \tag{10.76}$$

where $u_{\vec{k}}(\vec{r})$ is a periodic function in the lattice. Upon introducing random disorder into the system, the wave number entering into the Ioffe–Regel condition is from (10.76) given by

$$k = |\vec{k}\,| \tag{10.77}$$

where the wavevectors of the Block modes are restricted to the first Brillouin zone. This entails a representation of the solutions in which the wave functions and their eigenvalues are presented in the reduced zone scheme. In the reduced zone scheme, all the wave functions are labeled by wavevectors in the first Brillouin zone, and band indices are added to distinguish between different solutions with the same wavevector. This gives a complete characterization of all the solutions in terms of the form in (10.76).

These arguments carry over to the discussion of the localization of light in periodic optical systems in the presence of an additional random disorder. Specifically, the wave functions in a photonic crystal without disorder are of a Block form, similar to that in (10.76), and in the presence of the additional disorder the wavevectors of the Block form then, through (10.77) and (10.75), enter into the localization condition. Again, in these discussions the solutions of the Block forms are all presented within the reduced zone scheme.

For all types of periodic systems represented in the reduced zone, only wavevectors in the first Brillouin zone enter into the Ioffe–Regel condition. Consequently, for a fixed mean free path l_0 it is found in this representation that the states near the edges of the first Brillouin zone tend to be localized first. As the disorder is increased the boundary of the region of localized modes generally slips closer to the center of the zone [12].

As an addition important point, it is found that in looking at the localization properties of a Helmholtz wave equation of a type commonly studied in optics and acoustics the situation tends to be a little different from that of the Schrödinger type of equations. This causes some difficulty in predicting the degree to which localization occurs within optic and acoustic materials which is not present in electron systems. For example, in optical systems the Helmholtz equation is of the form [12]

$$\left[\nabla^2 + \frac{\omega^2}{c^2}\varepsilon(\vec{r})\right]\psi(\vec{r}) = 0 \tag{10.78}$$

where $\psi(\vec{r})$ is the wave function, the dielectric function $\varepsilon(\vec{r})$ plays the part of a random potential describing the optical interaction with the medium, and ω^2 is the energy eigenvalue. Now the energy eigenvalue multiplies the disordered scattering potential. A consequence of this is that the effects on the system of the scattering medium can be greatly decreased at low frequencies (long wavelengths), while being enhanced at higher frequencies (short wavelengths). This type of effect can result in a limitation on the lower frequency limits of the boundary of the localized states. A limitation which is not present in Schrödinger systems.

At low frequencies and long wavelengths an important factor characterizing the scattering of the Helmholtz equations is the Rayleigh scattering limit [12]. It is associated with, for example, the explanation of the blue tint of the Earth's sky. Specifically, in three-dimensional scattering from random arrays of particulates the Helmholtz equation gives a scattering cross section dependent on the wavelength as λ^{-4}. In the blue sky effect the long wavelength red components of light are weakly scattered out of the light passing through the atmosphere due to their small scattering cross sections while the shorter wavelength blue light is strongly scattered due to their smaller wavelengths [12]. The general angle of the diffusely scattered blue light gives the atmosphere its blue coloration. It is interesting to note in this regard, that this type of scattering also explains the yellow coloration of the Sun which in fact from space appears to be white.

The feature of long wavelength scattering generalizes to the treatment of scattering in general dimensions so that for particulate scattering in d-dimensions the Rayleigh scattering cross section is dependent on the wavelength as $\lambda^{-(d+1)}$. These Rayleigh effects in particulate scattering enter to provide a limitation on the observation of localization effects at long wavelengths in photonic crystal materials in terms of the system geometries required and material compositions employed [12]. Specifically, from (10.78) it is found that as λ becomes large eventually the localization is lost. Specifically, the scattering features in the photonic crystal tend to become invisible at long wavelengths.

On the other extreme, at short wavelengths, the scattering is limited by the typical size of the features composing the disorder in the material. Specifically, as the frequency of light is increased (i.e., at short wavelengths) the limit of ray optics is approached. In the ray optics limit the optical system is described by classical ray optics and the cross section for mean free path scattering tends to saturate. In this regard, the scattering features develop as shadows within the medium, and as the scattering becomes classical the mean free path also saturates at a constant value l_0. From (10.41) and (10.78) it is then seen that localization is eventually lost for $\lambda \ll l_0$. In photonic crystals formed of cylinders and spheres, for example, the saturation of the mean free path in the presence of a disorder is highly dependent on these specific features present and their packing in the material.

In summary, from these discussions it is seen that for both the long and short wavelength limits the optical system fails to be localized. It now remains as a question whether an optical system can ever develop a proper ratio $\frac{l_0}{\lambda}$ so that localization occurs in the random medium. If localization exists it can only occur at intermediate wavelengths between the two discussed limits. This question has been affirmatively resolved for optical and acoustic systems by both experiment and theory. We now turn to a discussion of these results for specific optical systems. The discussions begin with a treatment of localization in a disorder medium composed as a layering of dielectric slabs.

10.2.3 Example of localization in one-dimensional systems

In the following some results are presented for the localization of light in a disordered one-dimensional optical medium composed as a random layering of dielectric slabs [16–20]. As a specific model, the focus of the discussions is on a photonic crystal-like system. In this layered geometry the slabs composing the array are all of thickness a and are alternately layered along the x-axis so that slabs of vacuum are located between slabs of random dielectric media. Consequently, if all of the dielectric slabs in the array are identical the layering forms a one-dimensional photonic crystal of vacuum–dielectric slabs. In the random system, however, each of the dielectric slabs is composed of a medium with a different randomly set dielectric constant.

For the discussion of localization in the random array, the electromagnetic modes of interest are plane wave based modes moving along the x-axis and having a normal incidence to the surfaces of the slabs [16–20]. Specific considerations are presented for modes with electric fields polarized parallel to the y-axis. This restriction on the polarization, however, is not a fundamental limitation on the study of the solutions at normal incidence. As discussed earlier, all modes of the disordered array are localized so that the goal of the study is to determine the localization length of the wave functions in terms of the parameters characterizing the disorder in the system. As we have seen earlier the localization length is fundamentally related to the transmission of light through large finite arrays of the layering.

To begin, consider the transmission of light through a random array of $2N - 1$ slabs studied in the limit as N becomes large [16]. The array is characterized by a statistical distribution of the dielectric constants of each slab of the array so that the system is defined only in terms of the statistical properties of the layering. Specifically, a precise definition of the dielectric function of one of the arrays in the statistical set is given by the following: we treat the set of systems in which the one-dimensional dielectric function is described by the form [16]

$$\varepsilon_{2i-1} = \varepsilon + \delta_{2i-1} \text{ in the region } 2(i-1)a < x < (2i-1)a \tag{10.79a}$$

$$\varepsilon_{2i} = 1 \text{ in the region } (2i-1)a < x < 2ia \tag{10.79b}$$

where $i = 1, \ldots, N$, so that ε_{2i-1} and ε_{2i} label the dielectric constants of the $(2i-1)^{th}$ and $(2i)^{th}$ slabs, respectively. In addition, in the regions $x < 0$ and $x > (2N-1)a$ is vacuum. The $\{\delta_{2i-1}\}$ in (10.79) are a set of random numbers uniformly distributed over the intervolve $[-\delta, \delta]$ where $0 < \delta \ll \varepsilon$, and the average dielectric of each of the $(2i-1)^{\text{th}}$ slabs is $\varepsilon > 0$.

For the treatment of the transmission through the array, an electromagnetic plane wave with an electric field polarized along the y-axis is incident on the slabs, traveling along the x-axis from infinity at the right of the array to infinity on the left of the array [16]. The solution of the electrodynamics problem defined in this way, then reduces to the determination of a transfer matrix which can be used to describe the motion of the incident electromagnetic wave through the layered media. This is done by studying the transmission of an electromagnetic plane wave passing through a single dielectric slab and repeatedly applying the single slab result to generate the total transmission through the array of slabs. In this regard, we first consider the general problem of the transmission through a single dielectric slab, considering the system shown schematically in figure 10.7.

10.2.3.1 Transmission through a single dielectric slab

The system represented in figure 10.7 is a single dielectric slab surrounded by vacuum where the right- and left-hand sides of the dielectric slab are located at x'' and x', respectively. Within the vacuum regions, the electric plane wave solutions to the right and left sides of the dielectric slab have, respectively, the general forms [16]

$$\vec{E} = \left[E_{1R}e^{i\omega\left(\frac{x}{c}-t\right)} + E_{1L}e^{-i\omega\left(\frac{x}{c}+t\right)}\right]\hat{j} \tag{10.80a}$$

$$\vec{E} = \left[E_{2R}e^{i\omega\left(\frac{x}{c}-t\right)} + E_{2L}e^{-i\omega\left(\frac{x}{c}+t\right)}\right]\hat{j} \tag{10.80b}$$

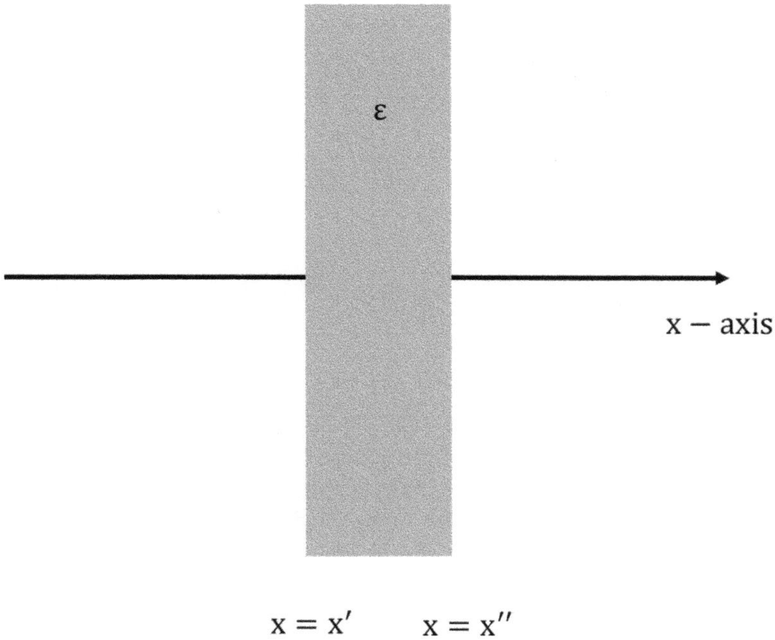

Figure 10.7. A single infinite dielectric slab surrounded by vacuum. The surfaces of the slabs are located at $x = x'$ and $x = x''$.

where 1 and 2 label the solutions in the left and the right regions of vacuum, and the subscripts R and L label the right and left moving components of these solutions.

By matching the electromagnetic solutions in (10.80) with those for the electric field within the dielectric slab (i.e., in the region $x' < x < x''$), it is found that [16]

$$\tilde{E}_2 = \underline{T}(x'', x'; \varepsilon)\tilde{E}_1 \tag{10.81}$$

where

$$\tilde{E}_1 = \begin{vmatrix} E_{1R} \\ E_{1L} \end{vmatrix} \tag{10.82a}$$

$$\tilde{E}_2 = \begin{vmatrix} E_{2R} \\ E_{2L} \end{vmatrix}. \tag{10.82b}$$

Here the 2×2 matrix relating the field amplitudes $\{E_{1R}, E_{1L}\}$ and $\{E_{2R}, E_{2L}\}$ are given by

$$\underline{T}(x'', x'; \varepsilon) = \begin{vmatrix} T_1(x'', x'; \varepsilon) & T_2(x'', x'; \varepsilon) \\ T_3(,x'',x'; \varepsilon) & T_4(x^{r},x'; \varepsilon) \end{vmatrix} \tag{10.83}$$

where

$$T_1(x'', x'; \varepsilon) = T_4^*(x'', x; \varepsilon) = \left\{ \frac{i}{2}\left[\sqrt{\varepsilon} + \frac{1}{\sqrt{\varepsilon}} \right] \sin\left[\sqrt{\varepsilon}\frac{\omega}{c}(x'' - x') \right] + \cos\left[\sqrt{\varepsilon}\frac{\omega}{c}(x'' - x') \right] \right\} e^{-i\frac{\omega}{c}(x''-x')}, \tag{10.84a}$$

$$T_2(x'', x'; \varepsilon) = T_3^*(x'', x'; \varepsilon) = \frac{i}{2}\left[\sqrt{\varepsilon} - \frac{1}{\sqrt{\varepsilon}} \right] \sin\left[\sqrt{\varepsilon}\frac{\omega}{c}(x'' - x') \right] e^{-i\frac{\omega}{c}(x'' + x')}, \tag{10.84b}$$

and ε is the dielectric constant of the medium forming the slab. The matrix $\underline{T}(x'', x'; \varepsilon)$ in (10.81) is known as the transfer matrix relating the solutions in the vacuum regions on either side of the dielectric slab to one another.

The transmission coefficients for an array of dielectric slabs is then obtained by determining the set of $\{x'', x'\}$ right the left positions of the surfaces of the dielectric slabs of the array and the values of the set $\{\varepsilon\}$ of slab dielectric constants corresponding to each of the slabs in the array [16]. From these measures of the system, the total 2×2 matrix relating the field amplitudes on the right and left sides of the array is composed as a product of matrices of the form in (10.81) and (10.83) for each of the dielectric slabs in the system. Using the fact that the field on the left of the array only propagates to the left, allows the transmitted and reflected wave amplitudes of the array to be written in terms of the amplitude of the wave incident on the system. The transmission and reflection coefficients of the system are ultimately determined in terms of these amplitudes.

10.2.3.2 Periodic photonic crystal solutions

In the case of a pure photonic crystal it follows that $\delta = 0$ in (10.79a) so that all of the dielectric slabs of the periodic vacuum–dielectric array are of dielectric constant ε. The transmission coefficient of the resulting system exhibits a band structure of stop and pass bands with a periodic dispersion relation given by the real solutions of [16]

$$\cos 2ka = \cos\sqrt{\varepsilon}\frac{\omega a}{c}\cos\frac{\omega a}{c} - \frac{1}{2}\left[\sqrt{\varepsilon} + \frac{1}{\sqrt{\varepsilon}} \right] \sin\sqrt{\varepsilon}\frac{\omega a}{c}\sin\frac{\omega a}{c}. \tag{10.85}$$

Here k is the effective wavevector of the one-dimensional wave function and ω is the modal frequency of the propagating modes in the array. A plot of the band structure obtained from (10.85) for a particular set of photonic crystal parameters is shown in figure 10.8.

10.2.3.3 Disordered systems: localization length

For the case of a disordered system it follows that $\delta \neq 0$ in (10.79a), so that now a set of dielectric configurations with proscribed statistical properties is studied. The boundary conditions applied to each of the set of solutions in the vacuum region to the right of the array has the form [16]

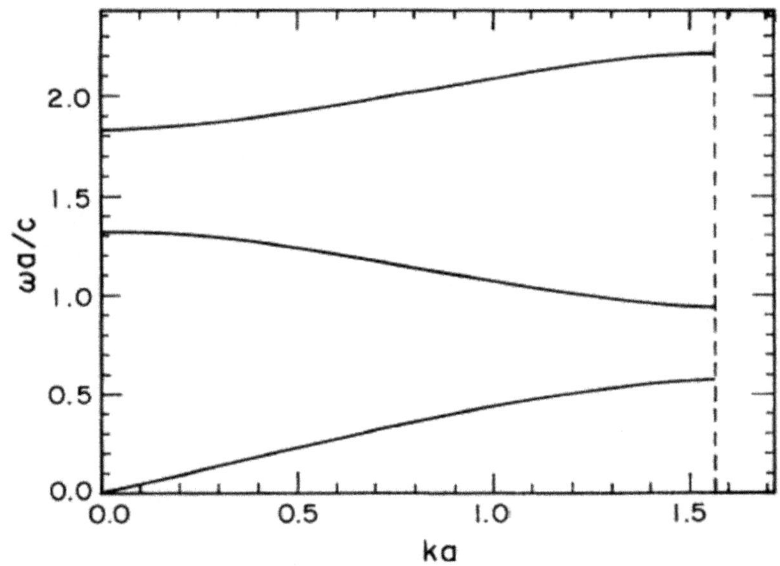

Figure 10.8. Dispersion relation for a one-dimensional photonic crystal with $\varepsilon = 9$. The plot is of frequency versus wave number for the three lowest frequency bands of the photonic crystal. See [16] for the details of the plot. Reprinted with permission from [16], Copyright 1993 American Physical Society.

$$\tilde{E}_{\text{Right Side}} = \begin{vmatrix} r \\ 1 \end{vmatrix} \tag{10.86a}$$

where the incident wave is of unit amplitude and the reflected wave has a reflected amplitude r, and the boundary conditions for each of these solutions in the vacuum region to the left of the array is the form

$$\tilde{E}_{\text{Left Side}} = \begin{vmatrix} 0 \\ t \end{vmatrix} \tag{10.86b}$$

where the transmitted wave has an amplitude t. (Note that the values of r and t will be different for different realizations of the array configurations.) In terms of the parameterization in (10.86) the transmission coefficient of the incident wave through the array is given by $T = |\, t \,|^2$, and a transmission coefficient is generated for each member of the set of statistical parameters characterizing the random system. In this way a set of transmission coefficients $\{T\}$ is obtained which is consistent with the statistical properties of the system being studied. The localization length of the model is then determined by a well-defined statistical averaging process involving the set $\{T\}$.

Specifically, it is presently shown that the localization length of the localized wave functions in the system is related to the set of generated transmission coefficients $\{T\}$ for all the statistical realizations of our model. The precise dependence of the localization length l of the model system on the set of T is found to be given by the form [16]

$$l = -\frac{L}{\langle \ln T \rangle}. \tag{10.87}$$

Here $L = (2N - 1)a$ is the length of the array and $\langle \ln T \rangle$ represents an average of $\ln T$ over all of the random configurations of the disordered system, i.e., over the set $\{T\}$.

To understand how (10.87) is arrived at in the optical system consider the following: the light incident on the disorder array of slabs couples to a mode of the slab. In the infinite array the mode would be localized because all modes in a one-dimensional disordered medium are localized. Consequently, at large distances from the maximum amplitude of the mode, the mode amplitude is expected to decay exponentially to zero. As a result, in a large array of slabs the energy flux incident on the right side of the random array should also decrease exponential with increasing array length, and the transmitted flux at the left side of the array should be an exponentially decreasing function of the length of the array.

The transmission coefficient of the array should then be of the form [16]

$$T = \alpha e^{-\frac{L}{l}} \tag{10.88a}$$

where $L \to \infty$ is the length of the array, α is a constant of proportionality, and l is a measure of the decay with length in the sample of the transmission. Consequently, in the limit of large L it follows that

$$\ln T = \ln \alpha - \frac{L}{l} \approx -\frac{L}{l}, \tag{10.88b}$$

and (10.87) is obtained upon averaging (10.88b) over the entire set of transmission coefficients $\{T\}$.

In figure 10.9 results are presented for the localization of a particular realization of the theory in (10.79) through (10.88). The plot studies the localization length as a function of frequency for a system with $N = 500$. The calculations are made by generating a large set of random configurations of dielectric slabs and using these to determine the localization length from the relation given in (10.87). For the details of the calculation the reader is referred to [16].

10.2.3.4 One-dimensional nonlinear optical media

The localization properties found in a randomly disordered one-dimensional system formed of materials with a Kerr type nonlinearity are significantly modified from those of a system formed solely of linear dielectric media [20, 21]. This has been shown from a number of analytical and numerical studies, and these are now discussed.

Earlier in the chapter it has been shown that in linear media systems the transmission through a finite array of dielectric layers is of an exponential decreasing form in the length of the array. This is given in (10.88). With the introduction of a random Kerr media the functional dependence of the localization on the length of the array is found to be fundamentally altered.

In this regard, the transmission properties of nonlinear systems of a finite number of layers is characterized by a power law decay in the length of the layering. This is given by [20, 21]

$$T = \alpha \mid L \mid^{-\beta}. \tag{10.89}$$

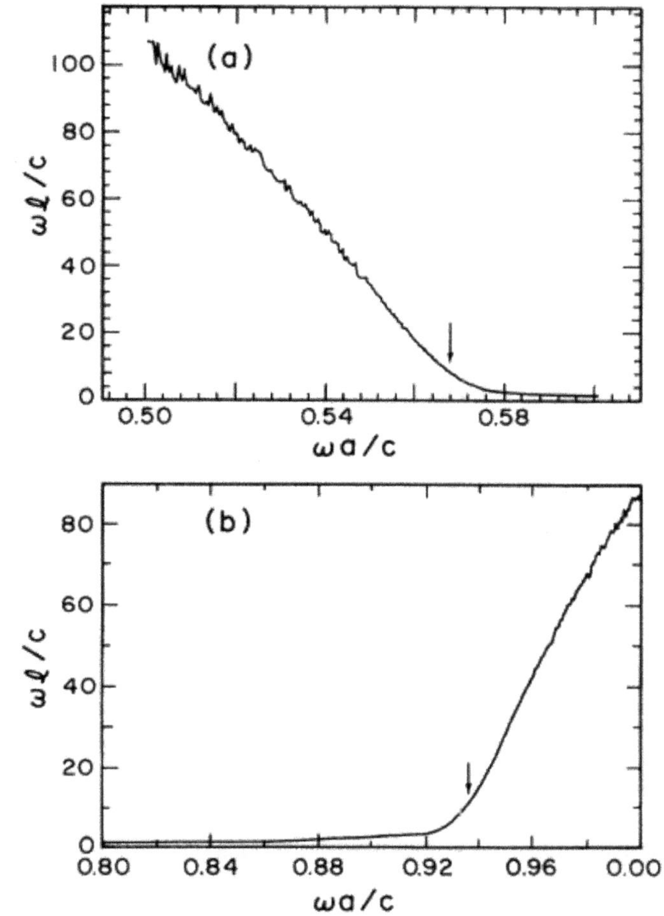

Figure 10.9. Plot of the localization length l versus the modal frequency ω for a system with $\varepsilon = 9$ and $\delta = 1$. (Note: in the limit that $\delta = 0$ the dispersion relation of the resulting pure photonic crystal is given in figure 10.8.) Two ranges of frequencies are examined, and the arrows in the plots indicate in: (a) the lower stop band edge at $\frac{\omega a}{c} = 0.5678$, and (b) the upper band edge at $\frac{\omega a}{c} = 0.9359$. The localization lengths are small for frequencies within the stop band regions and large for frequencies within the pass band regions. Reprinted with permission from [16], Copyright 1993 American Physical Society.

The nature of the exponent β has been investigated analytically [21] and by computer simulation in layered nonlinear media which is periodic on average [20]. From these studies it has been determined that the exponent β is within the region $0 < \beta < 2$.

10.2.3.5 Two-dimensional systems

For the study of localization in two-dimensional systems computer simulation methods are generally needed. In the following two different approaches to localization in two-dimensions are briefly discussed [22–26]. These include what is known as transverse localization and the related problem of strong localization in randomly rough two-dimensional dielectric arrays.

In the discussion of transverse localization [25, 26], a system of random dielectric cylinders is arrayed so that along the cylinder axes the array is translationally invariant, but perpendicular to the cylinder axes the system is randomly disordered. In a computer simulation study of this random system a semi-infinite array is formed by cutting the cylinders along a plane perpendicular to the cylinder axes and replacing the cylinders on one side of the cut by vacuum. A simulation is then run on the system for an incident beam of light with a circular cross section, incident normal from vacuum onto the cut surface of the random array. The incident beam is then normal incident on the plane perpendicular to the cylinder axes.

In the absence of disorder, the spot of light is found, as it moves in the direction parallel to the cylinder axes, to spread out indefinitely in the plane perpendicular to the cylinder axes. With the introduction of dielectric disorder in the plane perpendicular to the cylinder axes, however, a localization perpendicular to the cylinder axes is observed. Now as the light moves along the cylinder axes, the beam becomes localized to a region in the plane perpendicular to the cylinder axes, and this region of confinement in the plane is characterized by a finite localization length.

This phenomenon in the presence of disorder is known as transverse localization [25, 26] as the transverse spreading of the beam with distance along the cylinder axes is limited to a finite region of the plane transverse to the cylinder axes. The spreading in the transverse plane is significantly restricted from that of the beam in the absence of disorder. Again, the decay in the transverse plane is characteristically exponential in nature which is quite different from the beam spreading in the absence of disorder.

The second type of localization study involves the array of infinite cylinders extending between $\pm\infty$ along the cylinder axes so that there is complete translational symmetry along the cylinder axes [22, 23]. Into this array an infinitely long line antenna is places in the array along the cylinder axes which generates a radially expanding electromagnetic wave moving within the plane perpendicular to the cylinder axes. For disorder within the plane perpendicular to the cylinder axes, the electromagnetic fields from the antenna are confined within a region of localization about the antenna. The region of localization is characterized by a two-dimensional localization length which defines the radial exponential decay of the field intensity within the random medium. As in the case of transverse localization this confinement is distinct from the radial dependence of the field in the absence of disorder.

In figure 10.10 an example is given of the localization length is a specific two-dimensional disorder medium [23]. Specifically, a plot is made of the electric field amplitude as a function of radial separation from the line antenna for a disordered system. This is compared to the same antenna embedded within a photonic crystal without disorder. The rapid decay and confinement of the localized fields in the random system is evident. In this regard, the fields of the antenna within the photonic crystal are not confined by the ordered medium. For further details the reader is referred to the literature.

10.2.4 Weak localization effects in surface scattering

In this subsection a brief discussion is given of some of the properties of weak localization found in the elastic scattering of light from randomly disordered

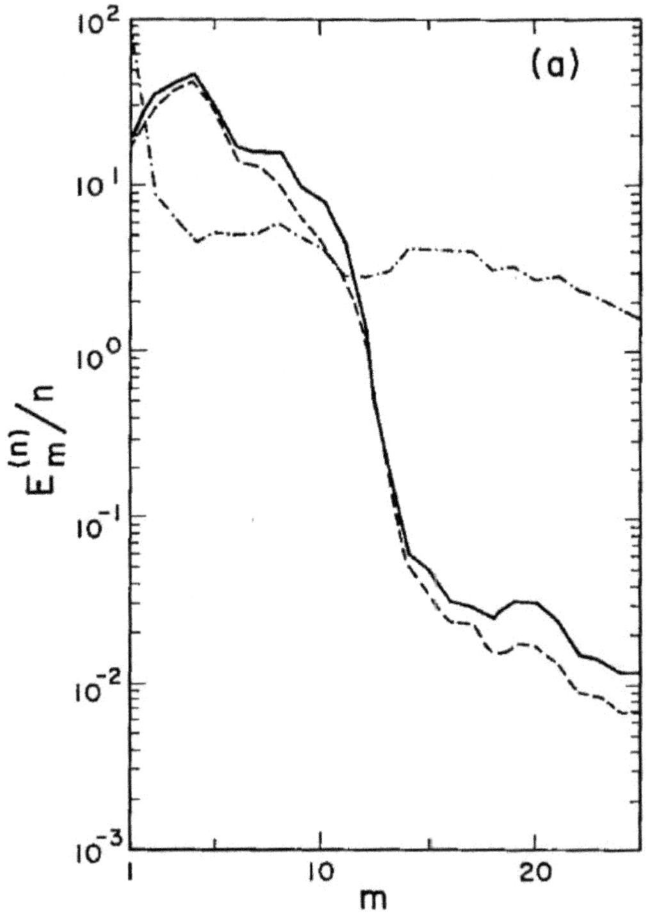

Figure 10.10. Plot of the electric field amplitude versus radial distance (measured in the dimensionless unit of length m). The antenna is located at the left side of the plot at $m = 0$. The solid and dashed curves are for the random system at two different frequencies. The dot–dashed curve is for an antenna within a photonic crystal formed with cylinders of the same dielectric constant ε. Notice that the amplitude decays more rapidly in the random medium. For the details of the system see [23]. Reprinted from [23], Copyright 1992, with permission from Elsevier.

surfaces [19]. Of specific interest are surfaces which support surface electromagnetic waves, sometimes referred to as surface plasmons or surface plasmon–polaritons. These types of surfaces are of interest as, in the presence of surface disorder, the surface electromagnetic waves they support display effects related to the phenomena of weak localization. Ultimately, an important property associated with the weak localization of the surface plasmons is observed as an enhancement of the diffusely scattered light from the rough surface. These phenomena involving elastic processes are a focus in the presentation to follow. Later after these discussions another set of effects related to surface enhanced Raman scattering (SER) and second harmonic generation will also be mentioned. The SER and second harmonic generation involves inelastic processes occurring at the surface and are distinct from the diffuse scattering considerations.

10.2.4.1 Surface waves

A common problem in the study of classical electrodynamics is the treatment of the refraction of a plane wave incident onto a planar interface. In this study the incident wave is found to separate at the surface between the two different optical media, splitting into a reflected and a refracted component. Each of these components is described by a Fresnel coefficient [1, 19]. The modal solutions for the scattering problem are characterized by the dielectric constants of the two media, the frequency of the incident wave, and the component of the wavevector projected in the surface. Since the system has complete translational symmetry in the planar interface the wavevector component parallel to the interface of the incident, reflected, and refracted waves are the same.

It is less commonly pointed out in electrodynamics texts that under the proper dielectric conditions it is also possible for there to be another set of modes associated with the planar interface. This additional set of classical electromagnetic solutions has the form of electromagnetic surface waves which are bound to and propagate along the interface. For these waves the field amplitudes of the electromagnetic waves decrease with large spatial separation from the interface. Due to the translational symmetry of the planar interface, the surface electromagnetic waves are also characterized as modes of wavevector along the planar interface. These surface electromagnetic modes are distinct from the refractive modal solutions and need to be included with them in order to form a complete set of modal solutions associated with the electrodynamics of the planar interface. It is interesting to consider how solutions of the refracted wave problem are related to those of the surface waves at the same vacuum–dielectric planar interface. This is now considered and ultimately related to a discussion of how surface disorder couples these two types of modes and in the process generates a diffuse scattering of light from surface disorder.

10.2.4.2 Components of the wavevectors in the plane of the surface

First consider the standard solutions of the problem of refraction at the interface. At a planar vacuum–dielectric interface the incident plane wave-based modes from vacuum onto the dielectric medium with real dielectric constant $\varepsilon(\omega)$ have wavevector components parallel to the interface given by the form [19]

$$k_{\|} = \frac{\omega}{c}\sin\theta_{\mathrm{inc}}. \tag{10.90}$$

Here ω is the frequency of the wave and θ_{inc} is the angle of incidence measured from the normal to the interface. The modes for the refracted waves in the system are, consequently, found from (10.90) to have $k_{\|} \leqslant \frac{\omega}{c}$.

Now make a similar consideration of the surface wave modes on the interface. These modes are obtained by looking at solutions in the vacuum and in the dielectric which decay in amplitude with increasing separation from the surface of the interface. In this way it is found, from matching the electromagnetic solutions above and below the planar interface, that the surface waves must have magnetic fields polarized parallel to the surface and perpendicular to the direction of propagation. It is also found that for the surface waves to exist, the dielectric constant needs to satisfy $\varepsilon(\omega) < 0$.

Under these conditions, the dispersion relation of the surface electromagnetic modes on the interface are given by [9, 19]

$$k_{sp} = k_{\parallel} = \frac{\omega}{c}\sqrt{\frac{\varepsilon(\omega)}{\varepsilon(\omega) + 1}}. \tag{10.91}$$

where k_{sp} is the surface electromagnetic wavevector which lies within the planar interface. From (10.91) it is seen that real solutions to (10.91) only exist in the case that

$$\frac{\varepsilon(\omega)}{\varepsilon(\omega) + 1} \geqslant 0. \tag{10.92}$$

which puts a further restriction on the dielectric constant that $\varepsilon(\omega) < -1$. As a consequence of these conditions [9, 19]

$$\sqrt{\frac{\varepsilon(\omega)}{\varepsilon(\omega) + 1}} \geqslant 1 \tag{10.93}$$

for surface electromagnetic waves to be supported on the interface.

In a comparison of (10.90) and (10.91) through (10.93), it is found that for a fixed frequency ω

$$\sin\theta_{\text{inc}} \leqslant 1 \leqslant \sqrt{\frac{\varepsilon(\omega)}{\varepsilon(\omega) + 1}}, \tag{10.94}$$

and the wavevector of the surface wave on the interface needs to satisfy $k_{\parallel} \geqslant \frac{\omega}{c}$. Consequently, at the frequency ω the refracting modes incident from vacuum are characterized by $k_{\parallel}$ which are smaller than the $k_{\parallel}$ of the surface electromagnetic waves. These two solutions are therefore distinct from one another so that there is no coupling between them on the flat surface. They may, however, become coupled to one another with the introduction of surface disorder.

10.2.4.3 Surface disorder and diffuse scattering of light

The lack of coupling between the two modes of frequency ω is changed with the introduction of surface disorder. On a disordered surface the translational symmetry along the interface is removed so that the wavevector $k_{\parallel}$ is no longer a constant of the motion for the problem. As a result, a scattering is introduced between the incident waves of frequency ω and the surface electromagnetic waves of frequency ω. Consequence, plane waves incident on the rough surface can now scatter into surface waves on the interface, and these surface waves can in turn be scattered by the rough surface into diffusely refracted and reflected waves traveling away from the surface.

In addition to the coupling of the two types of solutions, the phase coherent elastic scattering of the surface waves as they move along the interface subjects them to an enhanced backscattering effect. Specifically, the phase coherence of the waves contributions to a scattering of the surface waves back into the direction from which they came. This phase coherence is precisely the weak localization enhancement which was discussed in the previous sections. In this way, the weak localization of the surface waves ultimately shows up as an enhancement in the diffuse scattering of plane wave modes incident on the rough surface and diffusely scattered back into the incident wave.

The theory of this phase coherent scattering has been worked out analytically for weakly rough surfaces and by computer simulation studies for strongly rough surfaces. In the following some of the analytical results for weakly rough surfaces are presented. The reader is referred to the literature for the details of the treatment of the weak scattering limit and for the discussions of the computer simulations studies of the scattering from strongly rough systems [19].

10.2.4.4 Weak localization in the elastic scattering from rough surfaces

Consider a weak randomly rough vacuum–dielectric interface characterized by a surface geometry which on average is planar and by a homogeneous dielectric medium with dielectric properties that support surface electromagnetic waves[19]. (See figure 10.11(a) for a schematic representation.) Specifically, the dielectric constant $\varepsilon(\omega)$ of the medium is homogeneous and isotropic and fits the conditions discussed in (10.91) through (10.94) of the earlier subsection. In addition, the (x, y, z) coordinates characterizing the interface between the vacuum and dielectric medium are related by a surface profile function $x = \xi(y, z)$, assigning the x coordinate of the surface to the y and z space coordinates. On averaging over the statistical set of surface profiles $\langle\xi(y, z)\rangle = 0$, so that the surface is planar on average. In the following some of the basic properties of the electromagnetic surface waves and the diffusely scattered light from the interface are summarized.

As a further simplification to the initial discussions, begin by look at the one-dimensionally rough surface described by a surface profile function defined by $x = \xi(z)$. Consequently, in this model the surface is disordered along the z-direction but has translational symmetry along the y-axis. In the treatment to follow, the average plane of the interface, obtained upon averaging over the surface roughness, is the y–z plane. In this regard, of a particular focus are the localization of surface electromagnetic waves propagating in the z-direction, and the localization effects found in the diffuse scattering of light incident in the x–z plane from vacuum on the dielectric. This one-dimensional system provides a model which is simple enough to handle in the format of a perturbation theory treatment and which provides an illustration of the various physical aspects of the problem as found in more complex systems. For the details of the treatment to follow the reader is referred to the literature [19], the results of which are now discussed.

10.2.4.5 Surface waves

For surface waves propagating along the $z-$ axis, the spatial dependence of the electric field is described by a general plane wave form given by [19]

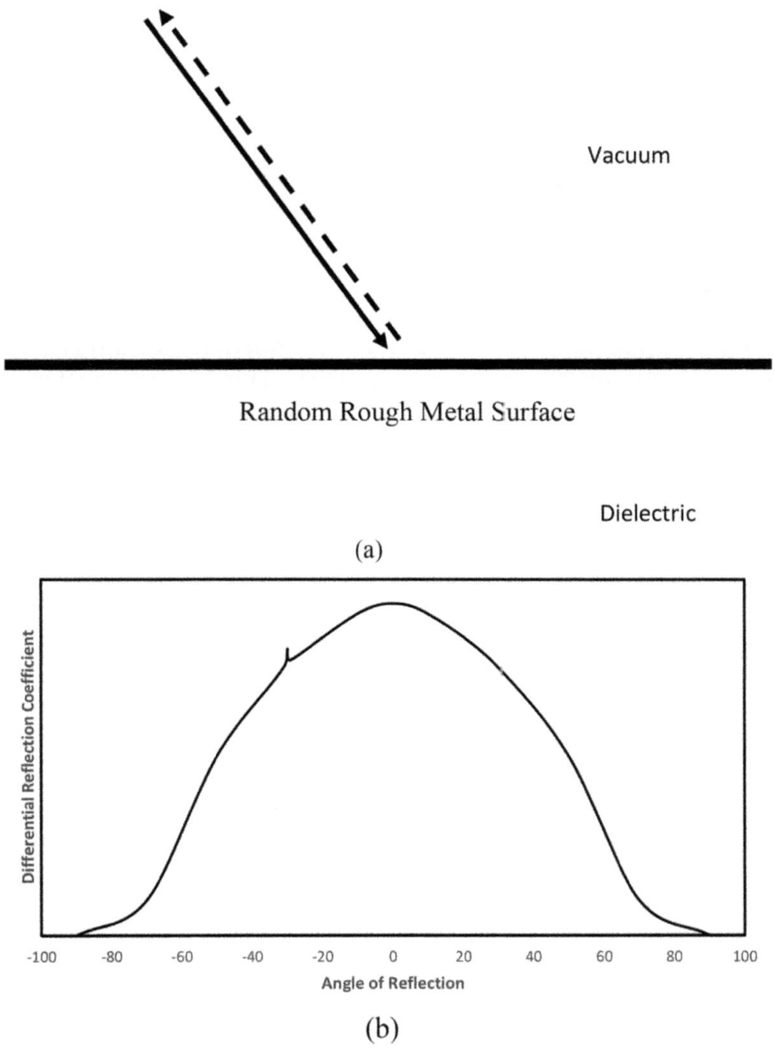

Figure 10.11. Schematic of a plane wave of light incident on a random rough vacuum–dielectric surface supporting surface electromagnetic waves. (a) The light is incident from vacuum (above the surface) in the direction of the solid arrow. The retroreflected light, indicated by the dashed arrow, is in the direction opposite the incident light. The mean surface is in the y–z plane. (b) A schematic illustrating a plot of the differential reflection coefficient of diffusely scattered light versus the angle of reflection for an incident plane wave with an incident angle of 32°. A peak in the intensity of diffusely reflected light is found within a small angular width centered about the retroreflection angle.

$$\vec{E}(x, y, z) = \vec{E}_0(x, y)e^{i(kz-\omega t)}, \quad (10.95a)$$

where

$$k = k_{sp} + ik_I. \quad (10.95b)$$

Here $k_{sp} = \frac{\omega}{c}\sqrt{\frac{Re[\varepsilon(\omega)]}{Re[\varepsilon(\omega)]+1}}$ is the real wave number of the surface wave, $k_I = \frac{1}{l_d}$ characterizes the decay in the surface wave amplitude, and l_d is a characteristic decay length along the interface arising from surface roughness scattering and from internal dielectric losses. In this regard, the scattering losses come from scattering of the surface wave into bulk electromagnetic waves propagating away from the surface and scatterings into other surface wave modes. Additional losses in the system arise from the imaginary part of the dielectric constant of the medium which accounts for molecular loss processes in the dielectric medium. Weak localization effects are also present in the characterization of the surface wave motion along the interface, but these are most noticeable in the consideration of diffuse scattering of bulk waves incident on the random interface. This diffuse scattering is now considered.

10.2.4.6 Scattering from the rough surface

For the problem of a plane wave of light traveling in the x–z plane and incident on the surface from vacuum, a general diffuse scattering is observed above the dielectric surface. The diffuse elastic scattering (i.e., no frequency change occurs.) is found as light propagating in the x–z plane away from the interface at general reflection angles measured from the normal to the mean surface. In the case of weak surface roughness, in addition to the diffuse scattering there is a component of a specular reflected wave from the mean surface which is reduced in intensity from that for the specular reflection of a perfect planar surface [19].

From a diagrammatic perturbation theory treatment of this problem a general analytic form is found characterizing the scattering from the interface in the limit of weak surface roughness. In this approach the intensity of radiation incident on the surface, with an incident angle θ_i as measure from the normal to the mean surface, is related to the intensity of diffusely scattered radiation from the surface within an infinitesimal angular interval $d\theta_r$ centered about a reflection angle θ_r. Again the reflection angle θ_r is measured from the normal to the mean surface. From these fluxes of radiation is computed the differential reflection coefficient. Specifically, the differential reflection coefficient is obtained from the ratio of the power reflected into a region of differential reflection angle $d\theta_r$ centered about the angle of reflection θ_r divided by the power incident on the surface at the angle of incidence θ_i.

As an additional important point in the discussions of diffuse scattering from the rough surface, it is necessary to consider the polarization of the incident radiation. This separates into two polarization types known as p- and s- polarizations. For the case of p-polarized light the magnetic field of the light is perpendicular to the plane of incidence while s-polarization light has the magnetic field of light polarized in the plane of incidence. In the study of localization effects, these two polarizations of light exhibit completely different properties. We begin by characterizing the properties of p-polarized scattering.

For the case of one-dimensional surface disorder, the differential reflection coefficient for light polarized with its magnetic field perpendicular to the x–z plane is found to have the general form given by [19]

$$\frac{\partial R}{\partial \theta_r} = A(\theta_r, \theta_i) + B(\theta_r, \theta_i)\frac{1}{\frac{\omega^2}{c^2}(\sin\theta_r + \sin\theta_i)^2 + 4k_I^2} \tag{10.96}$$

where $A(\theta_r, \theta_i)$ and $B(\theta_r, \theta_i)$ are slowly varying functions of the angles θ_r and θ_i. The first term, $A(\theta_r, \theta_i)$, in equation (10.96) contains the leading order Born scattering term of light from the surface roughness and some higher order process terms known as ladder diagrams. To leading order, the Born scattering term accounts for single scattering process from the rough interface, i.e., the light only encounters the surface once. The ladder process involves incoherent multiple scattering processes from the surface roughness. The second term in (10.96) arises from phase coherent multiple scattering processes discussed earlier in (10.65) through (10.73). Weak localization effects enter the scattering problem through the second term in (10.96), and their leading order occurrence accounts for a minimum of at least two scattering encounters with the interface roughness. In this regard, the processes present in the weak localization terms may be thought of as follows: the incident wave encounters the rough surface, excites a surface wave, and the surface wave is finally scattered into a wave propagating away from the surface. Higher order scattering process involve the repeated scattering of the surface wave with the surface roughness before its final scattering into an outgoing diffusely reflected wave from the interface.

Notice as an important distinction, it is found in the case in which the incident light is polarized with its magnetic field in the x–z plane the second term in (10.96) is absent. In this case no weak localization effects are observed in the diffuse scattering. The reason for the absence is that this polarization of the incident light does not couple to the localized surface waves. Consequently, with the absence of a coupling to the surface waves there can be no characterization generated in the diffuse scattering which signatures the presence of the localized surface waves. In the following, the discussions will continue for the polarization of incident light exhibiting localization effects, i.e., incident p-polarized light.

Of particular interest in the second term in (10.96) is the Lorentzian factor given by [19]

$$\frac{1}{(\sin\theta_r + \sin\theta_i)^2 + 4\frac{c^2}{\omega^2}k_I^2} = \frac{1}{(\sin\theta_r + \sin\theta_i)^2 + 4\left(\frac{1}{2\pi}\frac{\lambda}{l_d}\right)^2}. \tag{10.97}$$

This form accounts for the phase coherence of scattering processes coming from the presence of time-reversed paths in the multiple scattering events. From these processes arises an important angular dependence in the diffuse scattering which is solely the result of the presence of weak localization phenomena. These types of effects on the angular dependence of diffuse scattering from weakly rough surfaces can become important at optical and infrared frequencies.

A schematic representation of typical diffuse scattering results obtained from surfaces displaying weak localization effects are shown in figure 10.11(b). Here the enhanced backscattering is found to ride on an intensity of diffuse scattering which varies slowly with the angle of reflection, being observed as a small enhancement peak in the retroreflection direction. The retroreflection peak width is intimately related to the decay length and the localization length of the surface electromagnetic waves on the interface.

The schematic in figure 10.11(b) is meant only to give the reader an idea of the general effects found, and the reader is referred to [19] for detailed discussions of the diffuse scattering results for specific well characterized disordered surfaces. However, some general analytic discussions yielding simple expressions for the backscattering peak width and its dependence on surface dielectric properties and disorder are of interest. These are now presented.

10.2.4.7 Peak width

Specifically of interest is the case of (10.96) and (10.97) in which 4 $\frac{c^2}{\omega^2}k_I^2 = 4\left(\frac{1}{2\pi}\frac{\lambda}{l_d}\right)^2 \ll 1$. To see how this comes about consider the form of (10.97) for the limit in which $4\left(\frac{1}{2\pi}\frac{\lambda}{l_d}\right)^2 \ll 1$. This limit can often be easily achieved on weakly rough interfaces that support surface waves. In this limit, due to the presence of the $\sin\theta_r + \sin\theta_i$ term in the denominator of the Lorentzian form in equation (10.97), the Lorentzian is expected to be peaked within a narrow angular region centered about $\theta_r = -\theta_i$. Consequently, expanding $\sin\theta_r$ about $-\sin\theta_i$ (or equivalently taking $\theta_r = -\theta_i + \delta\theta_r$ for $\delta\theta_r \ll 1$) it follows that [19]

$$\sin\theta_r + \sin\theta_i = \delta\theta_r \cos\theta_i. \tag{10.98}$$

In this approximation, from (10.96) we find

$$\frac{\partial R}{\partial\theta_r} \approx A(\theta_r, \theta_i) + \frac{c^2}{\omega^2}\left(\frac{1}{\cos\theta_i}\right)^2 B(\theta_i, \theta_i)\frac{1}{(\theta_r + \theta_i)^2 + 4\left(\frac{1}{2\pi}\frac{\lambda}{l_d}\frac{1}{\cos\theta_i}\right)^2} \tag{10.99}$$

where $\theta_r + \theta_i = \delta\theta_r$ and the Lorentzian term in (10.99) is only significant for $\theta_r \approx -\theta_i$. In addition to these angular considerations, the Born term in $A(\theta_r, \theta_i)$ involves single scattering processes while the lowest order processes in the $B(\theta_i, \theta_i)$ term involve two scatterings. Consequently, the $A(\theta_r, \theta_i)$ term makes a larger contribution than the $B(\theta_i, \theta_i)$, except near $\theta_r \approx -\theta_i$.

While the first, Born, term in (10.96) and (10.99) remains slowly changing in the variables θ_r and θ_i the second term is rapidly changing about $\theta_r \approx -\theta_i$. Specifically, the second term contributes a sharply peaked Lorentzian centered about $\theta_r = -\theta_i$ in the particular limit that

$$\frac{1}{\pi}\frac{\lambda}{l_d}\frac{1}{\cos\theta_i} \ll 1. \tag{10.100}$$

This requires, however, that the wavelength of the plane wave radiation incident on the surface be much shorter that the decay length of the surface wave along the interface, i.e., $\lambda \ll l_d$. In addition, due to the presence of the $\cos\theta_i$ factor in the denominator, narrower peaks are found to occur for $\cos\theta_i \approx 1$ near normal incidence, and the peaks broaden out as the angle of incidence increases. Consequently, as the angle of incidence approaches $\theta_i \approx \frac{\pi}{2}$ the width of the peak becomes large and the backscattering enhancement is washed out.

From the Lorentzian form in (10.99) it follows that the full width of the diffuse reflection peak at half the maximum $\Delta\theta_r$ coming from the contribution of the second term is given by [19]

$$\Delta\theta_r = \frac{2}{\pi}\frac{\lambda}{l_d}\frac{1}{\cos\theta_i}. \tag{10.101}$$

In the case that $\cos\theta_i = 1$ it follows that the full width at half maximum of the backscattering peak is given in degrees by $\Delta\theta_r^{deg} = \frac{360}{\pi^2}\frac{\lambda}{l_d} \approx 36.48^\circ\frac{\lambda}{l_d}$ so that if $l_d = 10\lambda$ the full width at half maximum is 3.6°. This is not an unreasonable relationship between the mean free path and the wavelength of exciting waves in the study of plasmonic systems.

The enhancement effects discussed analytically here have been observed in the diffuse scattering from a number of weakly rough metal surfaces, and the reader is referred to the literature for discussions of these with comparisons made between theory and experiment [27]. Related to these studies a variety of computer simulation treatments have also confirmed the backscattering enhancement in the diffuse reflection from general rough surfaces. In addition, results on random weakly rough interfaces which combine periodicity and disorder [28] have also been treated as well as scattering associate with magnetic effects [29] and scattering on liquid metal surfaces involving the presence of surface ripplon disorder [30]. Weak localization effects on thin metal films have been shown to be present in both the light diffusely reflected and diffusely transmitted through thin films [31].

10.2.4.8 Additional mechanism of enhanced backscattering

As an addition side issue, the reader should be aware that backscattering enhancements of light can be present on randomly rough surfaces through an alternative mechanism to that of weak localization surface wave enhancements. One alternative mechanism known as the shadow effect is commonly observed in astronomy. Specifically, when the Earth comes between the Sun and a celestial body such as the Moon, the object can display a higher intensity of light back reflected to Earth per its angular resolution in the sky than at other Sun–Earth–object positionings. Consequently, the full moon looks brighter per area of the sky it occupies than does a quarter moon per area of the sky it occupies [19].

The shadow effect is a result of large surface roughness and has nothing to do with the presence of surface electromagnetic waves. To understand how this effect works, consider for example a flat area of sand randomly strewn with rocks, and upon which a plane wave of light is incident with an incident angle θ_i measured from the mean normal to the surface. Each of the rocks cast a shadow on the surface. In the case that the light diffusely reflected from the surface at a reflection angle $\theta_r = -\theta_i$ is studies, the dark patches of shadow on the surface from the rocks are covered by the light reflected back to the observer by the rocks themselves. On the other hand, however, if light reflected from the surface at $\theta_r \neq -\theta_i$ is studied, the observer sees various black shadow patches from the shadows put on the surface from the rocks. With the presence of black shadow patches the light intensity per surface area for an $\theta_r \neq -\theta_i$ observation is decreased from the light intensity per surface area viewed at $\theta_r = -\theta_i$. This is found to be the case for studies of planets and satellites and even clouds of particulates present in space. It is also found on geophysical features and vegetation on earth [19]. In the context of shadow casting, the apparent enhanced backscattering from these types of astronomical and geophysical systems is known as the opposition effect.

The shadowing effects exist on surfaces with large enough roughness that shadows are cast and do not require the presence of surface electromagnetic waves [19]. On rough surfaces which support surface electromagnetic waves, however, both shadowing effects and weak localization effects can contribute to enhanced backscattering. In this regard, only on weakly rough surfaces which do not cast shadows can one be assured that the enhanced backscattering is solely from a weak localization of surface waves.

The effect known as the optical glory is another backscattering enhancement, but this time it is found to occur in rain drops [19]. Consider a spherical rain drop with a plane wave incident on it. Treating the problem in ray optics: a ray incident on the left edge of the sphere is refracted, generates some virtual surfaces waves, and is reflected within the drop. By a combination of these processes some of the incidents ray energy leaves the drop at the right edge of the drop and travels anti-parallel to the original incident wave. Similarly, part of the incident wave hits the right edge of the drop and emerges at the left edge of the drop traveling anti-parallel to the incident ray. The two processes just outlined are time-reversed paths which add phase coherently to give an enhancement in the retroreflection direction. The drop has rotational symmetry so an infinite sequence of time-reversed paths are generated by rotation about the optical axis of the drop.

The optical glory is seen in the scattering of light from clouds. It is commonly observed as a bright ring about the shadow cast by an airplane on a bank of clouds as seen by an observer on the plane. In addition, certain colorations are generated due to the dispersion of light in water. This is again an effect pertaining to larger dimensional systems. The reader is referred to the literature for a discussion [19].

10.2.4.9 Second harmonic generation at a random rough surface

In addition to weak localization effects in the diffuse reflection of light elastically scattered from randomly rough surfaces, it is also possible to observe angular anomalies in the second harmonic components of light which may be generated above the random surface [32]. These anomalies occur in the diffusely generated second harmonics of light above the surface and arise from phase coherent weak localization processes of the electromagnetic surface waves generated on the interface.

As in the elastic scattering processes the anomalies develop from surface wave scattering on the rough surface, where now these scatterings involve both the fundamental and the second harmonic frequencies. The angular widths of the diffusely generated light anomalies are similarly related to the mean free paths of the surface electromagnetic waves as are those occurring in the case of elastic diffusely scattered light.

Specifically, a plane wave of light at frequency ω incident on such a surface generates a diffuse second harmonic at frequency 2ω which exhibit two angular anomalies [32]. The first anomaly is centered about the normal to the mean surface for 2ω radiation propagating away from the interface. A second harmonic anomaly is also generated about the backscatter direction of the ω radiation incident on the mean surface. These effects arise from weak localization in surface waves of frequencies ω and 2ω.

It should be noted in these discussions that the enhanced elastic backscattering at the rough surface is not affected by the presence of the very weak second harmonic effects in the system. The second harmonic processes are generally much smaller than those in the elastic scattering of light.

As a final note on inelastic effects, we mention another set of inelastic processes involving surface electromagnetic waves which are of technological interest. This is the technique of surface enhanced Raman spectroscopy. In these phenomena the strong fields created on surfaces by the excitation of surface electromagnetic waves are used in spectroscopic studies of molecules deposited on the surface. For discussions of these the reader is referred to the literature [1, 2, 19].

References

[1] Marder M P 2005 *Condensed Matter Physics* (New York: Wiley)

[2] Kittel C 1996 *Introduction to Solid State Physics* 8th edn (New York: Wiley) ch 18, 19 and 22

[3] McGurn A R 2002 Photonic crystal circuits: localized modes and waveguide couplers *Phys. Rev.* B **65** 075406

[4] Algul H G, Khazhinsky M, McGurn A R and Kapenga J 1995 Impurity modes from impurity clusters in photonic band structures *J. Phys.: Cond. Matter* **7** 447

[5] Algul H G, Chang Y F, Zhong Y and McGurn A R 1995 Impurity modes in 1-d photonic band structures *Physica* B **205** 19

[6] McGurn A R 2004 Resonant excitation of off-channel localized impurity modes by a photonic crystal waveguide *J. Phys.: Cond. Matter* **12** 5243

[7] Lee P A and Ramakrishnan T V 1985 Disordered electron systems *Rev. Mod. Phys.* **57** 287

[8] Segev M, Silverberg Y and Christodoulides D 2013 Anderson localization of light *Nature Photonics* **7** 197

[9] McGurn A 2020 *Introduction to Photonic and Phononic Crystals and Metamaterials* (San Rafael, CA: Morgan & Claypool)

[10] Mott N F 1968 Conduction in noncrystaline systems *Phil. Mag.* **17** 1259

[11] Mott N F and Davis E A 1979 *Electronic Processes in Non-Crystalline Materials* (Oxford: Oxford University Press)

[12] Shing P (ed) 1990 *Scattering and Localization of Classical Waves in Random Media* (Singapore: World Scientific)

[13] McGurn A R 2018 *Nanophotonics* (Cham: Springer)

[14] Cai W and Shalaev V 2009 *Optical Metamaterials* (Berlin: Springer)

[15] Bergmann G 1984 Weak localization *Thin Films Physics Reports* **107** 1

[16] McGurn A R, Maradudin A A, Christensen K and Mueller F 1993 Anderson localization in one-dimensional randomly disordered optical systems that are periodic on average *Phys. Rev.* B **47** 13120

[17] McGurn A R and Maradudin A A 1994 Localization of light in linear and nonlinear random layered dielectric systems that are periodic on average *Proc. SPIE* **2117** 2

[18] Freilikher V D, Liansky B A, Yurkevich I V, Maradudin A A and McGurn A R 1995 Enhanced transmission due to disorder *Phys. Rev.* E **51** 6301

[19] McGurn A R 1990 Enhanced retroreflectance effects in the reflection of light from randomly rough surfaces *Surf. Sci. Rep.* **10** 357

[20] McGurn A R and Maradudin A A 1994 Localization of light in linear and non-linear random layered dielectric systems that are periodic on average in The Proc. Third Int. Conf. on Elect. Trans. Opt. Prop. of Inhomogeneous Media *Phys.* A **207** 435

[21] Devillard P and Souillard B 1986 polynomial decaying transmission for the nonlinear schrodinger equation in a random medium *J. Stat. Phys.* **43** 423

[22] McGurn A R and Maradudin A A 1996 Perturbation theory results for the diffuse scattering of light from two-dimensional randomly rough surfaces *Waves Random Media* **6** 251

[23] McGurn A R, Sheng P and Maradudin 1992 Strong localization of light in two-dimensional disordered dielectric media *Opt. Commun.* **91** 175

[24] Pincemin F, Maradudin A A and McGurn A R 1994 The strong localization of light in a random two-dimensional dielectric medium that is periodic on average *Proc. SPIE* **2117** 12

[25] van Albada M P and Lagendijk A 1985 Observation of weak localization of light in a random medium *Phys. Rev. Lett.* **55** 2692

[26] de Raedt H, Lagemdijk A and de Vries P 1989 Transverse localization of light *Phys. Rev. Lett.* **62** 47

[27] Gu Z, Dummer R S, Maradudin A A and McGurn A R 1989 Experimental study of the opposition effect in the scattering of light from a randomly rough metal surface *Appl. Opt.* **28** 537

[28] Arsenieva A A, Maradudin A A, Lu J L and McGurn A R 1993 New features in the scattering of light from random surfaces that are periodic on average *Opt. Lett.* **18** 1588

[29] McGurn A R, Maradudin A A and Wallis R F 1991 Enhanced backscattering in a magnetic field, *Waves Random Media* **1** 43

[30] McGurn A R and Maradudin A A 1992 Enhanced backscattering of light from the surface of a liquid metal *Solids: a special volume in honor of the 65th birthday of Prof. Minko Balkanski* ed J L Birman, C Sebenne and R F Wallis (Amsterdam: North Holland)

[31] McGurn A R and Maradudin A A 1989 An analogue of enhanced backscattering in the transmission of light through a thin film with a randomly rough surface, *Opt. Commun.* **72** 279

[32] McGurn A R, Leskova T A and Agranovich V M 1991 Weak localization effects in the generation of second harmonics of light at a randomly rough vacuum-metal grating *Phys. Rev.* B **44** 11441

www.ingramcontent.com/pod-product-compliance
Ingram Content Group UK Ltd.
Pitfield, Milton Keynes, MK11 3LW, UK
UKHW051856100926
8556IPUK00005B/647